UPLC-MS/MS METHODS FOR DIFFERENT

TYPES OF CATEGORY DRUGS

:: Author ::

Dr. Darshan V. Chaudhary

PUBLISHED BY

The New Era International Publishing House
H.Q. At & Po. Chaveli., Ta- Chansma,
Dist- Patan, North Gujarat, India, Asia.
www.iphouseindia.com

First Publication: 22nd November, 2015

Copyright: Author
(c) *Dr. Darshan V. Chaudhary*

ISBN:- 978-1-51947-232-8

Price: Rs.800/- INDIA
$ 15 OUTSIDE INDIA

PUBLISHED BY

The New Era International Publishing House
H.Q. At & Po. Chaveli., Ta- Chansma,
Dist- Patan, North Gujarat, India, Asia.
www.iphouseindia.com

PREFACE

Man's use of drug and medicines dates back to the earliest times of civilization. Important human necessities such as the cure to a disease and precautions towards it has serendipitously took advantage of new discoveries in drug at the same time, rapid technological developments are now stimulating the chemistry and pharma industries to embrace technology, a trend strengthened by concerns regarding health, energy, raw materials, and the environment. Important advances in our understanding of the nature of drugs and their action were made in the late 19th and early 20th centuries, seeding the explosive expansion from the 1950s and 60s onward to the present billion dollar pharmaceutical based industries.

Much of the information about drugs has been made possible because they have functions largely because of their mode of action. Life is a dynamic process that involves constant changes in chemical composition. These changes are regulated by catalytic reactions, which are regulated by biological activities involving wide across the body. Few medical specialists' and allied biologists continue to think of this as a simple task, but we know that life as we know it could not exist or a disease would not be cured or prevented successfully without the use of targeted specific drugs whilst a proper due medication. Ideally, we would examine drug moieties within an intact cell, but this is difficult to dispense along with the targeted drug of interest. Consequently, drugs are studied in vitro after their formulation and delivery. It is not only one of thousands of drug working in concert within cells, but it is one but which readily demonstrates the main features.

Recent developments in the fields of medicine and drug chemistry are bringing ever more powerful means of analysis to bear on the study of drug structure and function that will undoubtedly leading to the rational modifications to match specific requirements and also the design of new drug with novel and multifunctional or a broad spectrum properties. The drug market and number of competitive drug discovery based processes is growing rapidly, because of cheaper production methods, new application fields, and new drugs from both synthetic as well as natural sources. The possibility to dramatically change drug properties by directed evolution and efficient methods to screen for new discovery in the field of medicine, makes it feasible to use that are specifically tailored to their application and process conditions. We expect that drug discovery technology is close to a major breakthrough, owing to many factors ranging from simple cost savings, the strongly increasing demand for chiral chemicals, the trend towards sustainable pharmaceutical development (more accessibility, quick targeted response and an adequately evident mode of action), and last, but not least, the opportunities created by emerging technologies.

Being treating radically different from the traditional ways, short and concise but complete, the subject of the presented book attest to be a practical instrument of work for students, technicians, young researchers or even experimented researchers because as the author said "health is today's most precious commodity" and why not, for any person interested in understanding of how nature works. Drug discovery is a helpful, innovative resource for practicing researchers in the chemical, pharmaceutical, and food science industries.

The development and validation of bioanalytical assay methods suitable for quantitation of the selected drugs (Alosetron, Amoxicillin and Clavulanic acid, Clarithromycin, Gliclazide, Lercanidipine) in biological matrices is discussed in this work. Relevant literature sources were consulted to understand the different parameters that must be included in method development and validation, to identify what constitutes a good assay method and to know the international regulations pertaining to bioanalytical methodology that determine whether a developed assay method is acceptable or not. Further, literature search is done to collect information on assay methods reported for the selected drugs (discussed in respective chapters). The different aspects of these assay methods viz. extraction, instrumentation, total turn-around time and sensitivity, selectivity etc. were assessed. Thus, an objective was set to develop selective, sensitive and rapid UPLC-MS/MS assay methods that have short and simple extraction procedures, consume small amounts of solvent and biological fluid for extraction and rapid compared to existing methods in the literature. Systematic validation as per USFDA guidelines is done for all the developed methods. The parameters investigated include selectivity, sensitivity, cross specificity, carry over effect, linearity, accuracy and precision, absolute and relative recovery, absolute and relative matrix effect, stability in plasma and dilution integrity. The application of these methods for bioequivalence study is conducted with test and reference formulation of the selected drugs on healthy human subjects. The pharmacokinetic parameters investigated include Cmax, Tmax, t1/2, AUC and Kel. Assay reproducibility was demonstrated by incurred samples reanalysis to ensure that the bioanalytical methods developed as per standard guidelines are rugged and reproducible for pharmacokinetic and bioequivalence studies. This study is conducted strictly in accordance with the guidelines laid down by International Conference on Harmonization and USFDA.

In present book, chapter 1 introduces the scope of bioanalysis of selected drug viz. Alosetron, Amoxicillin Clavulanic acid, Clarithromycin, gliclazide, and Lercanidipine. Chapter 2 describes the Experimental part of the study. Chapter 3 elobarates analysis, properties, bioequivalence and applications of Alosetron, while chapter 4 emphasis on analysis, properties bioequivalence and applications of combination drugs including Amoxicillin and Clavulanic acid. Chapter 5 discusses chemical properties, bioequivalence and applications of Clarithromycin. Chapter 6 elaborates scope, bioequivalence and applications of gliclazide and the last chapter 7 emphases on analysis, properties bioequivalence and applications of Lercanidipine.

The contents of the book will be useful to the students of Biotechnology, Industrial Chemists, Pharmaceutical science and technology, Food science and technology, Public health sciences etc.

I express my heartfelt thanks to Dr. Pranav Shrivastav, Professor, Chemistry Department, Gujarat University, Ahmedabad, India, for his constant guidance across my PhD. Work and without the same platform I would not be able to compile this book. I am very thankful to my other colleague contributors, Dr. Daxesh Patel, Jaivik Shah, Priyanka Shah and Edvin Pithawala for critical evaluation of each chapter across the book. I would like to express my gratitude to my family members especially to my parents for their love affection and care and last but not the least to my beloved Reema for her everlasting love, motivation and sacrifice for the time taken in compiling this book.

I am grateful to publisher for their concern, efforts and encouragement, especially for their excellent cooperation in the task of preparing and publishing this book.

*- **Darshan Chaudhary***

TABLE OF CONTENTS

CHAPTER-1: *Introduction*

CHAPTER-2: *Experimental – Materials, Instrumentation and Methodology*

CHAPTER-3: *Analysis of* **amoxicillin and clavulanic acid** *by UPLC-MS/MS in human plasma for pharmacokinetic application*

CHAPTER-4: *New improved UPLC-MS/MS method for reliable determination of* **clarithromycin** *in human plasma to support a bioequivalence study*

CHAPTER-5: *Sensitive and rapid determination of* **gliclazide** *in human plasma by UPLC-MS/MS and its application to a bioequivalence study*

CHAPTER – 1

INTRODUCTION

☞*CONTENTS*

1.1 THE PERSPECTIVE

1.2 THE AIM

1.3 BACKDROP 1FOR THE TITLE

    1.3.1 Sample preparation techniques

    1.3.2 High performance liquid chromatography

    1.3.3 Ultra performance liquid chromatography

    1.3.4 Mass spectrometry

    1.3.5 Liquid chromatography - mass spectrometry

1.4 BIOANALYTICAL METHODOLOGY

1.5 BIOANALYTICAL METHOD VALIDATION

1.6 BIOAVAILABILITY AND BIOEQUIVALENCE

1.7 BRIEF OUTLINE ABOUT THE WORK UNDERTAKEN

1.8 REFERENCES

1.1 The Perspective

A drug, is any substance that, when absorbed into the body of a living organism, alters normal bodily function. In pharmacology, a drug is "a chemical substance used in the treatment, cure, prevention or diagnosis of disease or used to enhance physical or mental well-being" [1]. The number of drugs introduced into the market is increasing every year. These drugs may be either new entities or partial structural modification of the existing one. Basically, for different type of diseases, drugs belonging to different categories are used.

The rapid development of pharmaceuticals has brought about a revolution in improving human health. Guided by pharmacology and clinical sciences, and driven by chemistry, pharmaceutical research in the past has played a crucial role in the progress of development of pharmaceuticals. In the field of pharmaceutical research, the analytical investigation of bulk drug materials, intermediates, drug products, drug formulations, impurities and degradation products and biological samples containing the drugs and their metabolites is very decisive [2].

The process in drug development starts with the innovation of a drug molecule that shows therapeutic value to battle, control, check or cure diseases. The synthesis and characterization of such molecules which are also called active pharmaceutical ingredients (APIs) and their analysis to create preliminary safety and therapeutic efficacy data are prerequisites to identification of drug candidates for further detailed investigations [3].

Drug bioanalysis, employed for the quantitative determination of drugs in biological fluids, plays a significant role in the evaluation and interpretation of bioequivalence, pharmacokineticand toxicokinetic studies. It is an integral part of characterization of drug from the time of its discovery and during various stages of drug development leading to its market authorization and thus development of sound bioanalytical method is of paramount importance. Drugs that are given in combination can produce effects that are greater than or less than the effect predicted from their individual potencies. Knowledge of drug levels in body fluids, such as whole blood, plasma, serum and urine, allows the optimization of pharmacotherapy and provides a basis for studies of patient compliance, bioavailability, pharmacokinetics and the influences of co-medications. Selective and sensitive analytical method development becomes necessary during the qualitative and quantitative analysis of drugs that are purported to display pharmacological activity, in determination of multiple drugs in combating a disease, biotransformation investigation, drug monitoring for therapeutic benefits and for *invitro* experiments [4-6].

Determination of drugs in a biological matrix is difficult in comparison to its formulations. Biological matrices like whole blood, plasma, serum and urine contain mainly water and other components like dissolved proteins, glucose, clotting factors, mineral ions, hormones and acids [7, 8]. These components may interfere at the time of quantification of analyte of interest if matrix free sample solution is not injected [9]. Drug absorption in body depends upon the properties of drugs and also some patient related factors, therefore, it is not always possible to avail high drug concentration in biological samples. Thus, efficient extraction procedures are imperative for successful bioanalysis of drug(s).

To make drugs serve their purpose, various chemical and instrumental methods have to be developed at regular intervals which are involved in the estimation of drugs. These pharmaceuticals may develop impurities at various stages of their development, transportation and storage which make their detection and quantitation to be extremely crucial after administration. In this regard, the analytical instrumentation and developed methods play an important role. Therefore it is imperative to provide an efficient detection and quantification procedure by utilization of suitable analytical techniques.

From the commencement of official drug analysis, analytical assay methods were included in the compendial monographs with the aim to characterize the quality of bulk drug materials by setting limits of their active ingredient content. In recent years, the assay methods in the monographs include non-instrumental method (titrimetry) followed by instrumental methods like spectrometry, chromatography, capillary electrophoresis and some electroanalytical techniques which have been reported in the literature. Figure 1 highlights classification of various analytical techniques.

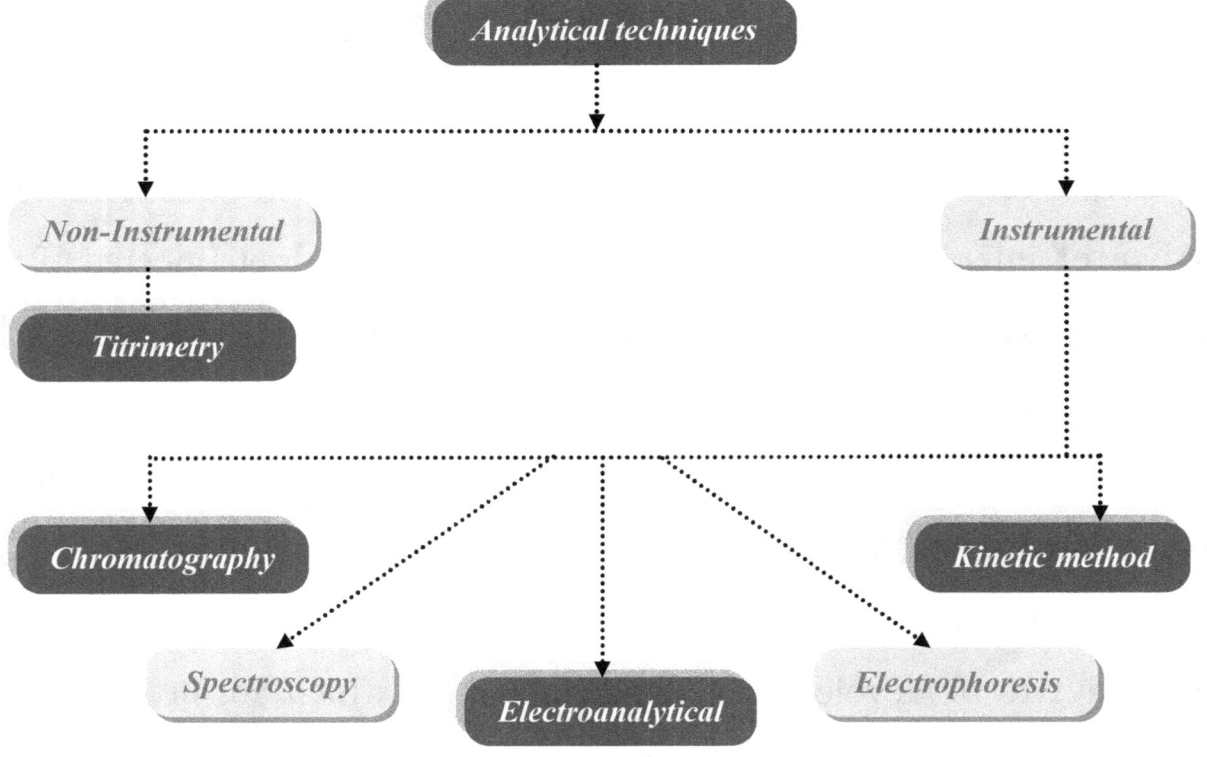

*Figure 1.***Classification of various analytical techniques**

Methods generally used in the analysis of drugs are radioimmunoassay (RIA), capillary electrophoresis (CE), gas chromatography (GC), GC-mass spectrometry (GC-MS), high performance liquid chromatography (HPLC) with UV, fluorescence, refractive index and mass spectrometric detection (MS) [10-12].LC-MS/MS has become an ideal and widely used method in the analysis of drugs and their metabolites due to its unmatched sensitivity, extraordinary selectivity and rapid rate of analysis [13]. Analytes that are easily separated by liquid chromatography can be detected even at lower concentration by MS/MS detection using different ionization techniques like electospray (ESI), atmospheric chemical ionization (APCI) and atmospheric pressure photo ionization (APPI) [14-17].

High performance liquid chromatography is the premier technique for chemical and pharmaceutical analysis with an ability to separate, analyze, and/or purify virtually any sample. The principle of separation of analytes is based on differences in relative rates of migration through the column arising from differential partitioning of the analytes between the stationary and the mobile phase. Reverse phase HPLC having hydrophobic stationary phase and polar mobile phase is generally used for the analysis of most of the compounds [18]. Sample preparation plays an important role in achieving the desired selectivity and sensitivity in the analysis. It is necessary to clean the biological sample as much as possible to get matrix free sample solution. An efficient extraction procedure need to be developed that can give quantitative and reproducible recovery.

Sometimes, concentrating the sample after extraction, derivatization at sample processing step or at chromatographic stage and adduct ion formation can enhance the sensitivity of the method. Thus, development of selective and sensitive analytical methods for the quantitative evaluation of drugs and their metabolites are critical for the successful conduct of preclinical and clinical pharmacology studies [19, 20]. These requirements are generally met with HPLC, especially if combined with an advanced detection technique such as mass spectrometry (MS). Now days, analysis time of biological samples can be decreased sharply using ultra performance liquid chromatography (UPLC), but the choice of an appropriate sample preparation method is essential for reliability and accuracy of the analysis as separation of analytes from other matrix components on column takes a short time.

1.2 The Aim

1. To develop and validate high throughput, sensitive and rugged bioanalytical methods   for routine sample analyses based on LC-MS/MS detection.

2. Separation of the drug from biological matrices by solid phase extraction (SPE), liquid   liquid extraction (LLE) or protein precipitation (PP) technique. The extraction   procedure should employ smaller plasma volumes, with quantitative and precise recovery of the drugs.

3. The developed method should have the following merits:

    ✓ High selectivity and sensitivity

    ✓ High throughput

    ✓ Less time consuming and less laborious extraction procedure

✓ Lower sample (biological fluid) volume requirement for processing

✓ Quantitative and precise recovery

✓ Rugged and robust enough for routine analysis

4. Application of these methods for bioequivalence/bioavailability studies of the drugs in   healthy human volunteers.

In the present study, five different classes of drugs are selected for their bioanalytical method development, validation and their application to bioequivalence studies.

- *Alosetron* – Serotonin 5-HT$_3$ antagonist
- *Amoxicillin and clavulanic acid* – Antibiotic and β-lactamase inhibitorrespectively
- *Clarithromycin* –Antibiotic
- *Gliclazide*– Antidiabetic
- *Lercanidipine* – Calcium channel blocker

1.3 Backdropforthe Title

Chromatography, a physical method of separation in which the components/solutes to be separated are distributed between two phases, one of which is a stationary (stationary phase) while the other (the mobile phase) moves in a definite direction. It is an analytical tool widely employed for the separation, identification of chemical/pharmaceutical components in complex mixtures. The components must interact with the stationary phase to be retained and for selective separation. The mobile phase may be a gas, liquid or a supercritical fluid which moves over or through the stationary phase, carrying the components along with it. Mass spectrometer is generally used for quantification of compounds in different biological matrices/complex mixtures. Analysis of drugs using liquid chromatography-mass spectrometry is extensively used in pharmaceutical industry.

LC-MS assay selectivity can be readily achieved by three stages of separation of the analyte(s) of interest from unwanted components in the biological matrices as shown in *Figure 2*.

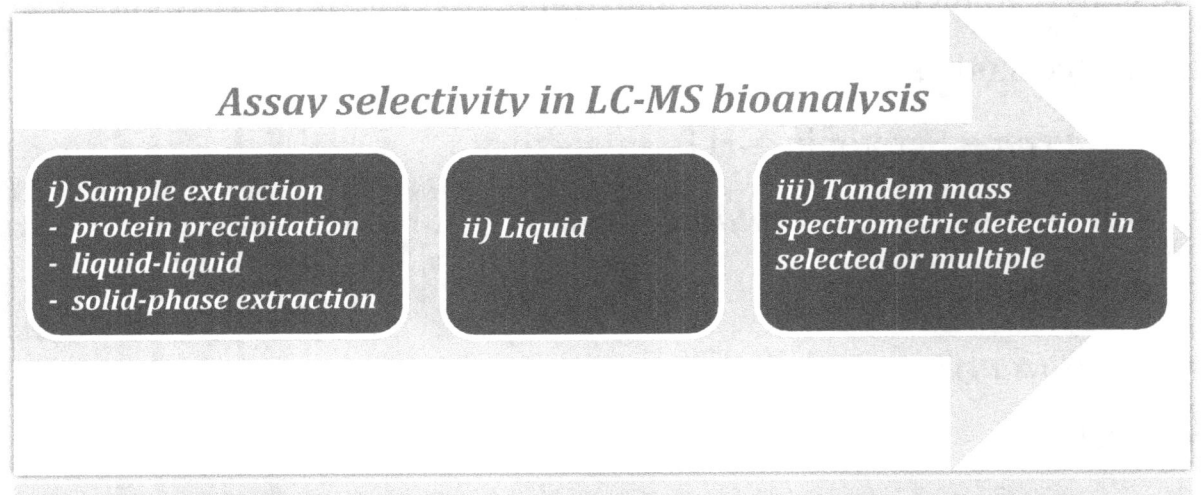

Figure 2. **Three stages for separation and analysis**

1.3.1 Sample preparation techniques

It is required to extract the drug from biological matrix beforeinjecting it into LC-MS/MS. Also, the sample clean-up plays an important role as it results in proper quantification of drugs. Due to sample clean-up, ion suppression or enhancement caused by endogenous impurities from biological fluids can be minimized. Protein precipitation, liquid-liquid extraction and solid phase extraction are the extraction methods generally used to extract the drugs from biological matrix*(Figure 3)*. The extraction method which gives higher and consistent recovery with minimum ion suppression/enhancement is used for analysis of incurred samples [21].

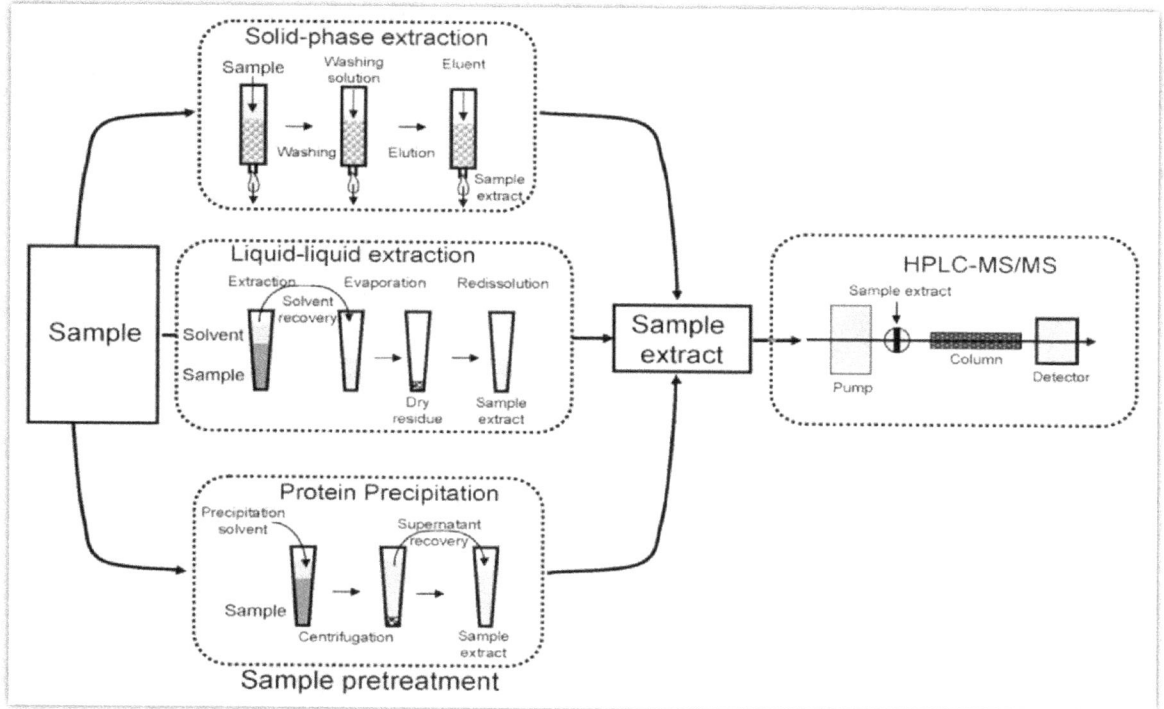

*Figure 3.***Principle strategies adopted in the determination of drugs in biological matrices**

i) Protein precipitationis a simplest procedure to remove proteins from a biological matrix. The inorganic/organic acid or an organic solvent such as perchloric acid (PCA), trichloroacetic acid (TCA), formic acid (FA), acetonitrile and methanol are used to precipitate proteins from biological fluids. The mixure is then centrifuged to removethe denatured proteins. After centrifugation, clear supernatant is injected directly or after drying and reconstitution into LC-MS/MS. It is a fast and cost effective extraction method but can give rise to significant matrix interferences which may then result in column clogging, ion suppression/enhancement and this require frequent system clean-up [22].

ii)Liquid-liquid extractionis a method used to separate compounds based on their relative solubility in two different immiscible liquids, usually water and an organic solvent. During extraction, the compound should be in unionized form and so pH adjustment of sample is necessary.

Sometimes it is required to back extract the analytes or multiple extractions to remove interferences from the sample. It is a cost effective method compared to solid phase extraction, but is tedious and time consuming as it requires drying followed by reconstition in appropriate solvent system. LLE is a simple and efficient method or the separation and concentration of relatively hydrophobic compounds. For some highly polar compounds, it is difficult to get matrix free samples using this extraction procedure [23].

iii)Solid phase extractionis an extraction method that uses a solid phase and a liquid phase to isolate one or more analyte(s) from the sample matrix. It is used to clean-up the sample before the chromatographic separation to quantitate the drug in the sample[24]. With SPE, many of the problems associated with LLE can be prevented, such as incomplete phase separations, less-than-quantitative recoveries, use of expensive, breakable specialty glassware, and disposal of large quantity of organic solvents. SPE methods are easy to perform, rapid and can be readily automated. Drugs can be extracted from small sample volumes with minimum use of solvents and reduced labour. In SPE, the sample is first loaded on a SPE cartridge, washed with suitable solvent to remove undesired components, followed by elution of desired analyte(s) into a collection tube. This method has a distinct advantage over protein precipitation as it affords clean and matrix free sample by washing out of undesired components. Different types of SPE cartridges are available to extract different types of drugs.

1.3.2Liquid chromatography (LC) [25, 26]

HPLC-based techniques have been a traditional mainstay of the pharmaceutical industry. It is a powerful technology that allows complex mixtures to be transformed into separated components. It is highly sensitive, reproducible, accessible, and well understood from an operator's standpoint. The output from the HPLC is its unique characteristic that distinguishes it from all other analytical techniques and the chromatogram, is defined and simple. Each peak is characteristic of a component; each chromatogram is diagnostic of an event or experiment associated with a drug development activity. When combined with the facts that nearly all compounds of pharmaceutical interest are amenable to HPLC methodologies and conditions and that critical information on nearly all events in the drug development cycle can be derived from HPLC chromatograms, it becomes evident why HPLC is a universally accepted analysis tool.

The various forms of chromatography employing a liquid rather than a gas as the moving phase are studied under the head 'liquid chromatography'. A useful classification of different LC techniques is based on the type of distribution mechanism applied in the separation. The four most widely studied LC methodologies include, partition (liquid-liquid), adsorption (solid-liquid), ion-exchange and size exclusion chromatography. Nowadays, HPLC is the most widely used analytical technique for the qualitative and quantitative analysis of pharmaceuticals, biomolecules, polymers, and other organic compounds. HPLC is a physical separation technique conducted in the liquid phase in which a sample is separated into its constituent components (or analytes) by distributing between the mobile phase (a continuous flowing liquid) and a stationary phase (sorbents packed inside a column). Modern HPLC uses high pressure to force solvent through closed columns containing very fine particles that give high-resolution separations. The column is the heart of the system. The efficiency of a packed column increases as the size of the stationary phase particles decreases, where the particle size ranges from 3-10 μm. In general, columns of 10 to 300 mm length are used with an inner diameter of 3 to 5 mm. A typical modular/integrated HPLC system consists of a multi-solvent delivery pump, an on-line degasser, an autosampler, a high pressure column, a column oven, a detector and a data-handling workstation. *Figure4* gives a schematic representation of a typical HPLC system.

Normal phase adsorption HPLC utilizes a polar stationary phase and less polar mobile phase. It is mainly used for the analysis of relatively nonpolar compounds. The retention of the components in a mixture increases with increasing polarity of the analytes. Silica gel is used as adsorbent in most applications, although alumina and chemically bonded stationary phases (with diol groups) are also used as well. The specific adsorption used in normal phase is the result of electrostatic forces between the permanent dipole of the silanol groups on the silica gel surface and the permanent or induced dipoles on the analyte molecules.

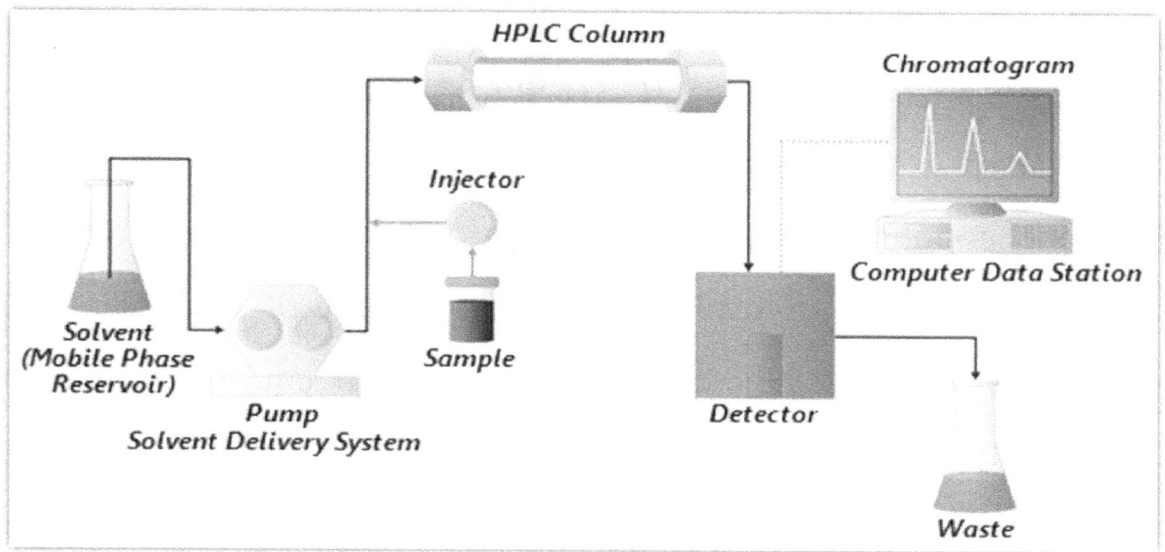

Figure 4.Schematic representation of a HPLC system

Reversed-phase system are used for most of the present LC applications, especially because the chemically bonded non-polar packing materials (stationary phase) are more easy to use. It is ideally suited for the analysis of polar and ionogenic analytes. In the reversed phase systems, the stationary phase is non polar or weakly polar and the mobile phase is more polar. Silica gel can be easily modified at the surface by chemical reactions with organochloro- or organoalkoxysilanes as shown in *Figure 5*.

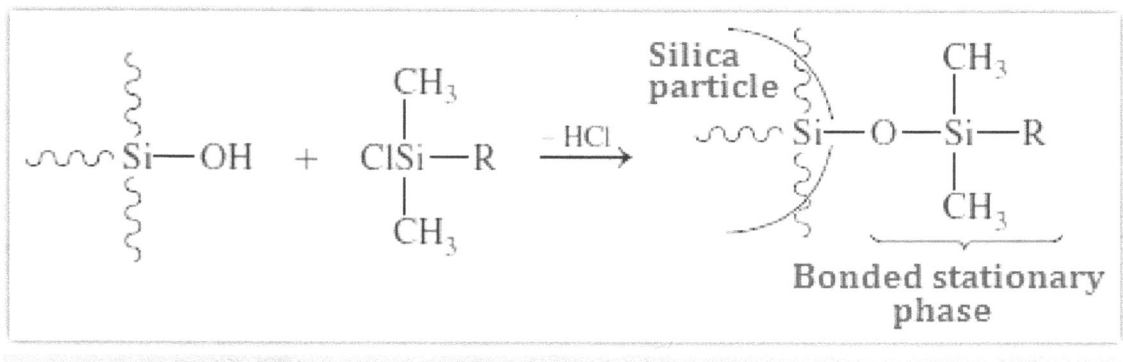

Figure 5.Modification in a silica gel

The siloxane bond is stable in organic and aqueous solvents in the pH range 2.5-8.0. Important R-groups that can be introduced in this way are,octyl (C_8), octadecyl (C_{18}), phenyl, *n*-propylamine, alkyldiol, alkyl-$N^+(CH_3)_3$ and phenylsulfonate (C_6H_4-SO_3^-). Non-polar, chemically modified silica gel or other non-polar packing materials, such as styrene-divinylbenzene copolymers, are used as stationary phases for aqueous-organic solvent mixtures up to pH-14 *(Figure6)*. Some of the advantages of bonded-phase materials are relative stability, short equilibration time, and their versatility [27, 28].

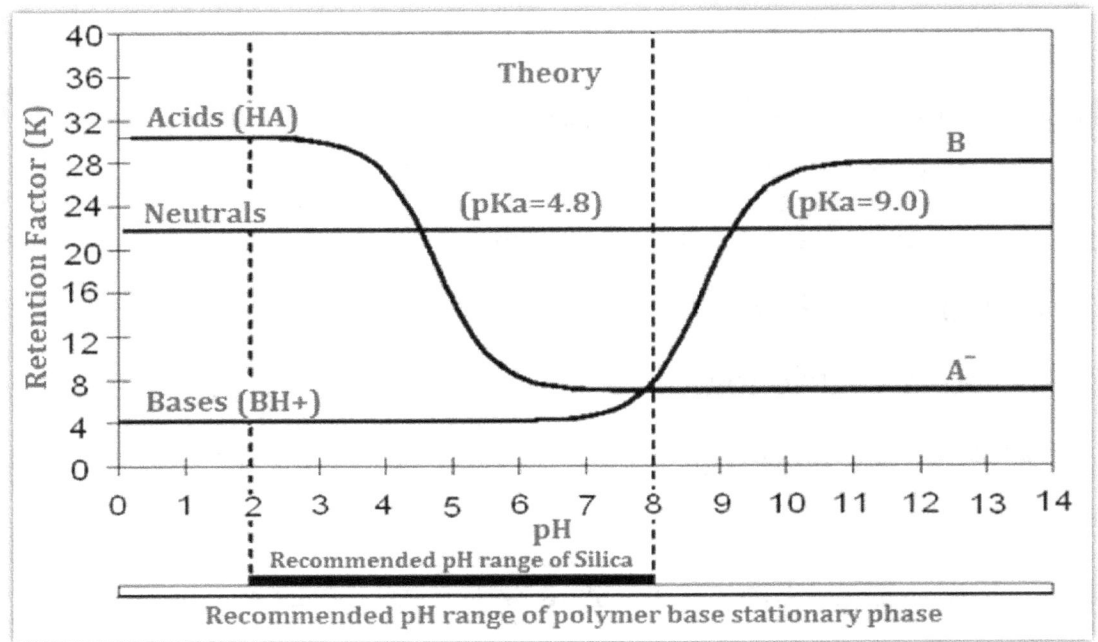

*Figure 6. **The retention capacity of acids and bases for the entire pH range***

Separation process in HPLC- The sample is injected by means of an injection port into the mobile phase stream delivered by the high-pressure pump and transported through the column where the separation takes place and the analytes are monitored with a flow-through detector like UV, fluorescence, photo diode array, refractive index, electrochemical or evaporative light-scattering.

The fundamental and important steps in creating highly selective phase systems include rigorous control of surface chemistry and adjustment of the final stationary phase properties by appropriate mobile phase selection, leading to specific solute-surface interaction and suppressing undesired side interactions.

Unlike in adsorption chromatography, the specific analyte-solvent interactions e.g. solubility effects, are most important in reversed-phase HPLC, since the interaction of the analyte and the bonded-phase material are weak and nonspecific like the Van der Waals interaction. The retention decreases with increasing polarity of the analyte. Mixtures of water or aqueous buffers and an organic modifier (methanol, acetonitrile etc.) are used as eluents. The percentage and type of organic modifier is the most important parameter in adjusting the retention of non-ionic analytes. Considerable attention is given to automated optimization of reversed-phase LC separations [29]. A buffer is frequently used in reversed-phase systems as it reduces the protolysis of ionogenic analytes, which in ionic form shows little retention. Phosphate buffers are widely applied for that purpose, since they span a wide pH range and show good buffer capacity. The use of buffer is obligatory in real world applications e.g. bioanalysis, where many of the matrix components are ionogenic.

In the analysis of acidic or basic compounds, adjustment of the pH is not always successful. The addition of an organic lipophilic ionic compound as a counter-ion for the ionic analytes results in the formation of ion-pairs that are well retained on the reversed-phase material. Traditionally, most pharmaceutical assays involve isocratic analysis, employing the same phase throughout the elution of the sample. Isocratic analyses are particularly common in quality control applications since they use simpler HPLC equipment and premixed mobile phase. In contrast, gradient analysis is suited for complex samples and those containing analytes of wide polarities in which the strength of the mobile phase (organic modifier content) is increased with time during sample elution. Hence the separation can be achieved in shorter time and with better efficiency.

Pharmaceutical separations can be divided into three categories: high throughput, high productivity and high resolution. These categories contain specific pharmaceutical applications, each of which has distinct separation goals. Traditionally, these goals have been achieved by utilizing conventional HPLC with typical column dimensions and particle sizes. But today pharmaceutical industry is facing major challenges of shortening the drug discovery and development time cycle to push new therapies into the market. With an increase in the number of drug candidates, the demand to improve the productivity of analyses in pharmaceutical development has significantly raised the level of interest in high-speed LC. The recent introduction of ultrahigh or ultra-performance liquid chromatography (UHPLC) has provided a new potential for method development and analysis.

1.3.3 Ultra performance liquid chromatography

Column technology with sub-2 μm particle size has revolutionized the field of separation science during the last decade. With the advent of smaller particle size material as stationary phase, there has been a renewed interest and a step function change in the way liquid chromatography is conducted today. UPLC with sub-2 μm particle size has demonstrated enhanced efficiency, superior resolution, higher sensitivity and much faster throughput compared to conventional HPLC with 3 or 5 μm particles. This has provided a tremendous boost for analyzing complex mixtures and streamlining extensive analytical workflow.

Liquid chromatography has stood the test of time in the field of separation science since its introduction in the early 1900s. The major breakthrough towards enhancing the performance of LC was in the 1960s when high pressure (500 psi, 35 bar) was used to generate flow through packed columns with smaller stationary phase particle size (≤ 10 μm). This led to a new era of 'high pressure liquid chromatography' (HPLC). Thereafter, with significant advancement in column technology it was possible to achieve even higher pressure (6000 psi, 400 bar) to give much superior separation performance and thus replacing the prefix 'high pressure' with 'high performance' in conventional LC. Since then different approaches have been adopted to improve chromatographic performance especially for analysis time, resolution and sensitivity. Although conventional HPLC with 3 or 5 μm particles find widespread use in environmental, clinical, toxicology and pharmaceutical analysis, it has relatively moderate efficiency and requires long analysis time for separating complex mixtures. Subsequently, several attempts were made to overcome these challenges by increasing the flow rates and reducing column lengths to enhance chromatographic efficiency [30]. Such modifications partially resolved the issues facing conventional HPLC, albeit with some limitations as they result in low phase ratio and small capacity factors. Another approach was also tried by increasing the column temperature, however it was largely ineffective especially for temperature sensitive compounds and could potentially damage the column material. Decreasing the particle size leads to significant increase in peak capacity and speed of analysis. ***Figure7*** shows the timeline for evolution of stationary phase particle size for LC.

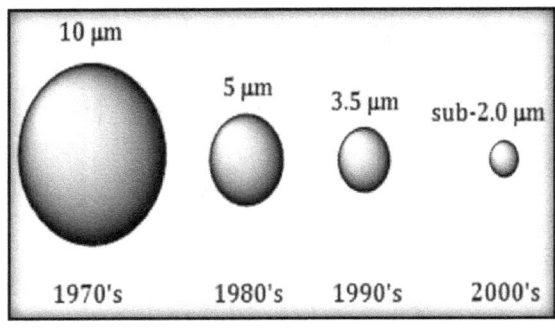

*Figure 7.**Evolution of stationary phase particle size in the past four decades***

One major drawback though of reduced particle size is that it can induce high back pressure as shown by the Darcy's law. This law governs the flow of liquids through packed columns and states that the flow rate is proportional to the pressure gradient.

is proportional to the pressure gradient. As the column back pressure varies inversely with the square of the particle size at constant linear flow rate [25], the operating pressure can go up to 15000-20000 psi which is difficult to handle with conventional HPLC instrumentation. This increase in the back pressure (> 10000 psi, ~700 bar) has led to this technique to be referred as high or ultrahigh pressure liquid chromatography (UHPLC or UPLC). Apparently, some practical hitches in working with smaller particles took almost 4 decades to realize its full potential. The year 2004 can be considered as the defining moment in LC when the first UHPLC system with sub-2 μm particle size was commercialized.

It has been proved that the chromatographic efficiency and resolution shows significant improvement with decreasing particle size. Nevertheless, to have the best chromatographic performance in terms of resolution, peak capacity, sharp peak shape for enhanced signal-to-noise ratio, lower carryover, less dispersion and shorter run times. Thus, in the present work to improve separation efficiency and shorten the analysis time, sensitive, selective and rapid UPLC method, in which column having 1.7 μm particle size has been employed for reliable measurement of selected drug candidates in human plasma.

Analysis by HPLC/UPLC-UV generally lacks sensitivity and specificity, especially in analysis of drug candidates in biological matrices. Thus, a need for more sensitive and selective detectors is paramount in such cases, where the drug is extensively metabolized and yet the retention time and UV spectral character remains the same as that of parent compound. Mass spectrometry detection and characterization is now a vital new tool in bioanalytical method development and pharmacokinetic analysis of drugs.

1.3.4 Mass Spectrometry

Mass spectrometry (MS) has progressed to become a powerful analytical tool for both quantitative and qualitative applications. Over the past decade, mass spectrometry has undergone tremendous technological improvements allowing for its application to proteins, peptides, carbohydrates, DNA, drugs, and many other biologically relevant molecules. Due to ionization sources such as electrospray ionization and matrix-assisted laser desorption/ ionization (MALDI), mass spectrometry has become an irreplaceable tool in the biological sciences. The MS principle consists of ionizing chemical compounds to generate charged molecules or molecule fragments and in the measurement of their mass-to-charge ratio.

The compounds can be ionized by different techniques like electron ionization (EI), chemical ionization (CI), electrospray ionization (ESI), atmospheric pressure chemical ionization (APCI), atmospheric pressure photon ionization (APPI), nano spray ionization (NSI), matrix assisted laser desorption ionization (MALDI), fast atom bombardment (FAB) and thermal ionization (TI). After ionization, ions are separated according to their mass-to-charge ratio by mass analyzer. Mass analyzers have electric and/or magnetic field in vacuum. Time of flight, quadrupole, quadrupole ion trap, linear quadrupole ion trap, and Fourier transform ion cyclotron resonance orbit trap are different mass analyzers that work based on different characteristics. Quadrupole is the most compatible analyzer as they are reasonably priced and make good multi-purpose instruments [26].

Tandem mass spectrometry (MS/MS) is more useful technique compared to normal mass spectrometry as it has more than one analyzer that gives better selectivity while analyzing the compounds and so nowadays it is the most powerful technique used in quantitative determination of compounds from biological fluids. First the mass analyzer separates the ions according to their mass-to-charge ratio from other ions. The ions of interest enter into the collision cell where they are fragmented by an inert gas (He, N_2 or Ar) using collision activated dissociation (CAD) process. The second mass analyzer allows selective fragmented ions (product ions) to be detected by the detector. Mass analyzing and detection process is carried out in high vacuum. The most important separation-in-space tandem mass spectrometer is the triple quadrupole as shown in *Figure 8*.

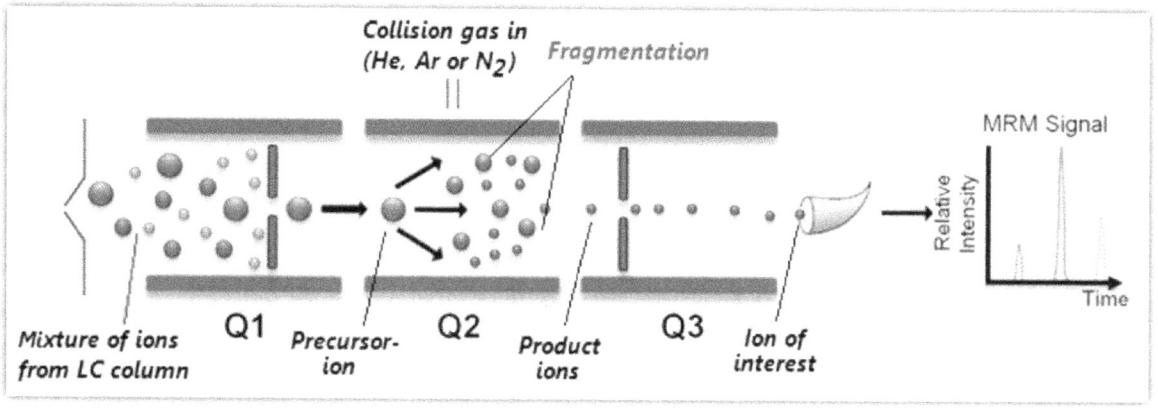

Figure 8. **A tandem mass spectrometry system**

1.3.5 HPLC- Mass spectrometry

High performance liquid chromatography (HPLC) coupled with mass spectrometry is an extremely powerful and indispensable methodology practiced in virtually every stage of pharmaceutical discovery and development process, including biological target discovery, biological assay for high throughput screening, characterization of physiochemical properties of drug candidates, drug metabolism and pharmacokinetics. Compounds are separated on column in HPLC and then enter into the mass spectrometer where they are first ionized in the source (parent ions). ESI, APCI and APPI are the ionization techniques mostly used for analysis of pharmaceutical compounds in biological fluids [31, 32]. A block diagram of liquid chromatography-mass spectrometer is depicted in *Figure 9*.

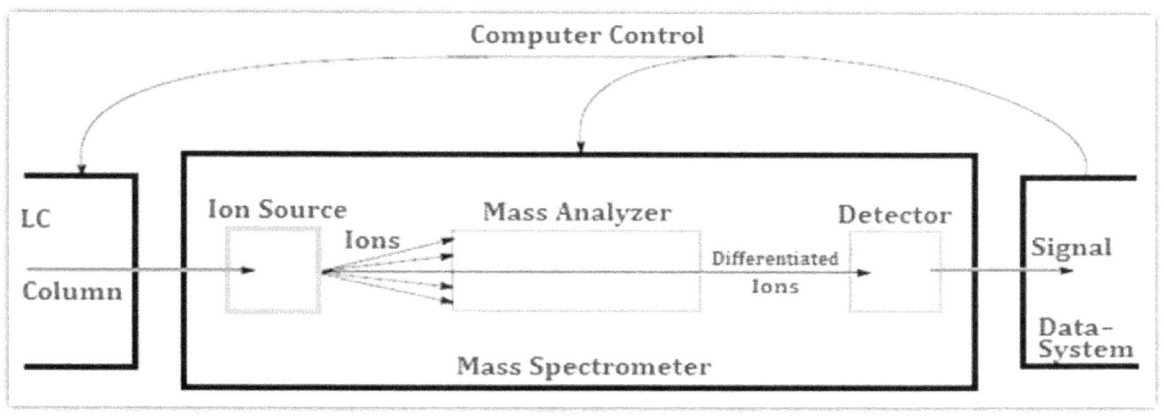

Figure 9. **Block diagram of liquid chromatography-mass spectrometer**

1.4 Bioanalytical Methodology

The process by which a specific bioanalytical method is developed, validated, and used in routine sample analysis can be divided into three elements, which are illustrated in *Figure 10*. The main analytical segments that comprise the bioanalytical methodology are method development, method validation and application in routine sample analysis.

Method development involves evaluation and optimization of the various stages of sample preparation, chromatographic separation, detection and quantitation. Initially, an extensive literature survey on the same or similar analyte is done followed by summarizing the main features of the work, which is of primary importance.

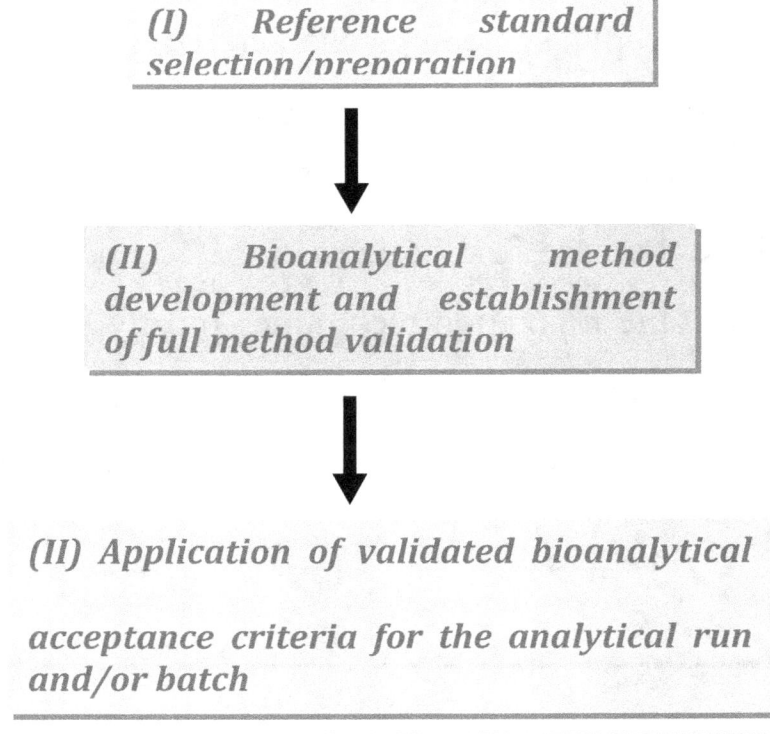

Figure 10. **General process for bioanalytical method**

Based on this information, the following selections could be made:

i) The choice of instrument that is suitable for the analysis of the analyte of interest. This includes the choice of the column associated with the instrument, the detector, the mobile phase in the HPLC, and the choice of carrier gas in GC.

ii) Choice of internal standard, which is best for the study. It must have similar chromatographic and ionization properties compared to the analyte.

iii) The choice of extraction procedure, which is quick and efficient, gives the highest possible recovery without interference at the elution time of the analyte of interest and has acceptable accuracy and precision which meets the intended study requirement.

Method performance is determined primarily by the quality of the procedure itself. The two factors that are most important in determining the quality of the method are selective recovery and standardization. Analytical recovery of a method refers to whether the analytical method in question provides response for the entire amount of analyte that is contained in a sample. Recovery is usually defined as the percentage of the reference material that is measured, to that which is added to a blank. This should not be confused with the test of matrix effect in which recovery is defined as the response measured from the matrix (e.g. plasma or blood) as a percentage of that measured from the pure solvent (e.g. water). Results of the experiment that compare matrix to pure solvent is referred to as relative recovery and true test of recovery is referred to as absolute recovery [21, 33].

In the presence of other sample components the ionization process can be changed and is mostly observed as a loss in response and referred to as matrix effect that can influence the accuracy, precision and lower limit of quantification of the assay. The matrix effect can occur with any biological matrix (plasma, serum, blood or urine), and it may vary depending on the source of matrix [34]. It is also compound dependent and it was shown that the methods based on electrospray ionization, mainly used for quantification of immunosuppressive drugs, are more likely to experience a matrix effect than that based on heated nebulizer [21]. Assessment of matrix effect is a very important step during method development/validation and cannot be omitted. There are two commonly used tests for evaluation of the matrix effect: post-extraction addition and post-column infusion [21]. The probability of the occurrence of the matrix effect can be reduced or eliminated by utilizing a more thorough clean-up procedure of the biological samples, modifying the chromatographic conditions, or allowing longer run times in order to enhance the chromatographic separation between the matrix-effect causing component and the analyte [34].

The choice of internal standard is also very important for minimizing the influence of matrix effect on accuracy of the assay. An isotopically labelled compound used as an internal standard will be affected by any matrix effect in the same way like analyzed drug so the accuracy in the quantitation should not be impaired provided that adequate signal from the drug and internal standard occurs so that signal to noise ratio is adequate for both.

1.5 Bioanalytical method validation

The search for the reliable range of a method and continuous application of this knowledge is called validation [35]. It can also be defined as the process of documenting that the method under consideration is suitable for its intended purpose [36]. Method validation involves all the procedures required to demonstrate that a particular method for quantitative determination of an analyte(s) in a particular biological matrix is reliable for the intended application [37]. Validation is also a proof of the repeatability, specificity and suitability of the method. Bioanalytical methods must be validated if the results are used to support the registration of a new drug or a new formulation of an existing one. Validation is required to demonstrate the performance of the method and reliability of analytical results [38]. If a bioanalytical method is claimed to be for quantitative biomedical application, then it is important to ensure that a minimum package of validation experiments has been conducted and that it yields satisfactory results [39].

Before discussing how to carry out the validation experiment, it is important to stress that validation in bioanalysis should not be considered as an isolated field. A consensus on common terminology for all analytical fields is therefore required. For the moment it is not yet possible to propose a validation terminology that is also in agreement with the recommendations of important international organizations such as the ISO (International Standard Organization), IUPAC (International Union of Pure and Applied Chemistry) and AOAC (Association of Official Analytical Chemists), since some differences exist between their documents [36].

For the validation of pharmaceutical drug formulations the discussion on a consensus terminology is relatively advanced. It is suggested to follow in general the proposal elaborated for the validation of drug formulation by the joint initiative of the pharmaceutical industry and the regulatory agencies of the three major regulatory authorities (the European Union, the USA and Japan) and the International Conference on

Harmonization (ICH) [40]. According to them the revised version of terminology to be included are bias (accuracy), precision, specificity, limit of detection, limit of quantitation, linearity, range and stability. The term stability is also specifically considered in the validation strategy for bioanalytical methods, which is prepared by the French group SFSTP (Societe Francaise des Sciences et Techniques Pharmaceutiques) [36].

On the other hand the guideline for industry by FDA [41] states that the fundamental parameters of validation parameters for a bioanalytical method validation are accuracy, precision, selectivity, sensitivity, reproducibility and stability. Typical method development and establishment for bioanalytical method includes determination of (1) sensitivity, (2) selectivity, (3) accuracy and precision, (4) carryover, (5) calibration curve, (5) recovery, (6) matrix effect, (7) dilution integrity and (8) stability. For a bioanalytical method to be considered valid, specific acceptance criteria should be set in advance and achieved for accuracy and precision for the validation of the QC samples.

1.6 Bioavailability and bioequivalence [42, 43]

The therapeutic effectiveness of a drug depends upon the ability of the dosage form to deliver the medicament to its site of action at a rate and amount sufficient to elicit the desired pharmacologic response. This attribute of the dosage form is referred to as physiologic availability, biologic availability or simply bioavailability. For most drugs, the pharmacologic response can be related directly to the plasma or blood levels. Thus, the term bioavailability is defined as the rate and extent (amount) of absorption of unchanged drug from its dosage form. Absolute bioavailability is the fraction of drug effectively absorbed after extravascular administration, when compared to the administration of the same drug intravenously. Relative bioavailability or bioequivalence between drugs, administered by the same extra vascular route, may be evaluated by comparing pharmacokinetic parameters related to bioavailability, i.e., to the quantity absorbed and to the rate of the absorption process. Bioequivalent drugs are pharmaceutical equivalents (same pharmaceutical formulation and quantity of the same active ingredient) that, when given in the same molar dose, in the same condition, does not present significant statistical differences regarding bioavailability. The rate or rapidity with which a drug is absorbed is an important consideration when a rapid onset of action is desired as in the treatment of acute conditions such as asthma attack, pain, etc. A slower absorption rate is however desired when the aim is to prolong the duration of action or to avoid the adverse effects. On the other hand, extent of absorption is of special significance in the treatment of chronic conditions like hyper-tension, epilepsy, etc.

If the size of the dose to be administered is same, then bioavailability of a drug from its dosage form depends upon 3 major factors:

i) Pharmaceutics factors related to physicochemical properties of the drug and   characteristics of the dosage form

ii) Patient related factors

iii) Route of administration

The influence of route of administration on drug's bioavailability is generally in the following order: parenteral > oral > rectal > topical with few exceptions. Within the parenteral route, intravenous injection of a drug results in 100% bioavailability as the absorption process is bypassed.

However, for reasons of stability and convenience, most drugs are administered orally. In such cases, the dose available to the patient, called as the bioavailable dose, is often less than the administered dose. The amount of drug that reaches the systemic circulation (i.e. extent of absorption) is called as systemic availability or simply availability. The term bioavailability fraction F (Bioavailable dose /Administered dose) refers to the fraction of administered dose that enters the systemic circulation.

In the relationship between dose and effectiveness or dose response, not only the amount of drug administered and the pharmacological effect of the drug are of importance but many other factors are responsible for the entry of a drug into the body. These factors are based on the physical and chemical properties of the drug substance and of the drug product. What happens to the active ingredient in the body after administration of a drug product in its various dosage forms? This entire cycle of processes is termed fate of drugs. Whether a blood/plasma level curve will reach its peak rapidly or slowly depends on the route of administration, the dosage form, the liberation rate of the drug from the dosage form, diffusion, penetration and permeation of the drug, its distribution within the body fluids and tissues, the type, amount and rate of biotransformation, recycling processes and elimination.

☞ *Plasma Drug Concentration-Time Profile*

A direct relationship exists between the concentration of drug at the biophase (site of action) and the concentration of drug in plasma. A typical plasma drug concentration-time curve obtained after a single oral dose of a drug and showing various pharmacokinetic and pharmacodynamic parameters is depicted in *Figure 11*. Such a profile can be obtained by measuring the concentration of drug in plasma samples taken at various intervals of time after administration of a dosage form and plotting the concentration of drug in plasma (Y-axis) versus the corresponding time at which the plasma sample was collected (X-axis).

The five important pharmacokinetic parameters that describe the plasma level-time curve and useful in assessing the bioavailabilities of a drug from its formulation are:

***Peak Plasma concentration (C_{max})*:** The point of maximum concentration of drug in plasma is called as the peak and the concentration of drug at peak is known as peak plasma concentration. It is also called as peak height concentration and maximum drug concentration. The peak level depends upon the administered dose and rate of absorption and elimination. The peak represents the point of time when absorption rate equals elimination rate of drug. The portion of curve to the left of peak represents absorption phase i.e. when the rate of absorption is greater than the rate of elimination. The section of curve to the right of peak generally represents elimination phase i.e. when the rate of elimination exceeds rate of absorption. Peak concentration is often related to the intensity of pharmacologic response and should ideally be above minimum effective concentration (MEC) but less than the maximum safe concentration (MSC).

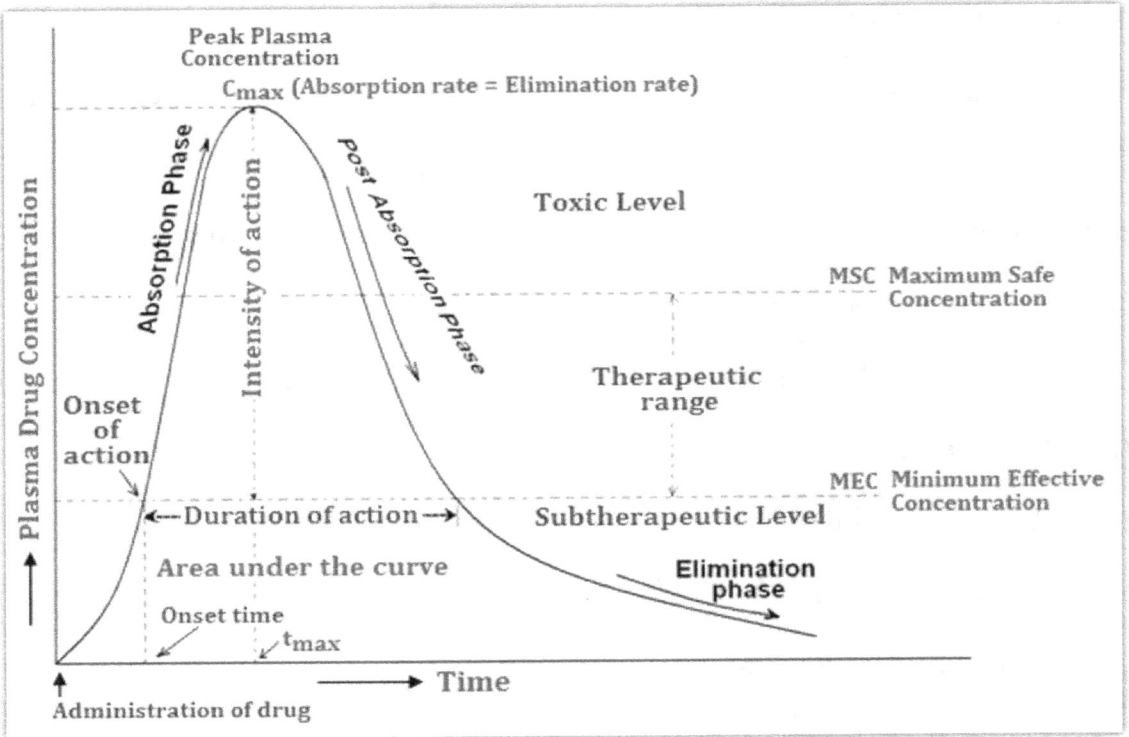

Figure 11. A typical plasma concentration-time profile showing pharmacokinetic-pharmacodynamic parameters

Time of peak concentration (T_{max}): The time for drug to reach peak concentration in plasma (after extravascular administration) is called as the time of peak concentration. It is expressed in hours and is useful in estimating the rate of absorption. Onset time and onset of action are dependent upon T_{max}. The parameter is of particular importance in assessing the efficacy of drugs used to treat acute conditions like pain and insomnia which can be treated by a single dose.

Area under the Curve (AUC): It represents the total integrated area under the plasma level-time profile and expresses the total amount of drug that comes into the systemic circulation after its administration. AUC is expressed in µg/mL × h. It is the most important parameter in evaluating the bioavailability of a drug from its dosage form as it represents the extent of absorption. AUC is also important for drugs that are administered repetitively for the treatment of chronic conditions like asthma or epilepsy.

Elimination half-life ($t_{1/2}$) and rate constant (K_{el}): The time required for the concentration of the drug to reach half of its original value is termed as elimination half-life. Elimination rate constant can be defined as the rate at which a drug is removed from the body.

☞ ***Incurred sample reanalysis***

Recently, the 'incurred' or study sample reanalysis (ISR) has become mandatory for bioanalytical methods used to support the drug development process. Viswanathan *et al.* [43] have suggested that an evaluation of the reproducibility in the analysis of incurred samples be performed on each species used for Good Laboratory Practices (GLP) toxicology assessments, as well as an appropriate evaluation of incurred sample reproducibility from clinical studies. Incurred or study samples can vary in their composition when compared with the standards and quality control samples used to validate the method and analyze these samples. During the 3rd American Association of Pharmaceutical Scientists (AAPS)/Food and Drug Administration (FDA) Bioanalytical Workshop, it was suggested that the reproducibility in the analysis of incurred samples be evaluated in addition to the usual pre-study validation activities performed. Although every attempt is made to formulate standards and QCs to be as similar to the study samples being analyzed as possible, "incurred" or study samples can differ in a variety of ways. These differences are dependent in part on whether the analyte(s) in question are small molecules or macromolecules. Moreover, it becomes even more important when metabolites are measured, as they may convert *in vitro* to their parent drug molecule [44]. In addition, the philosophy of ISR will ensure continuous review and monitoring of results through scientific procedures. This will help to adopt improved practice in bioanalysis, while safeguarding the optimal use of time, labour and laboratory resources [45, 46].

1.7 Brief outline about the work undertaken

The development and validation of bioanalytical assay methods suitable for quantitation of the selected drugs (Alosetron, Amoxicillin and Clavulanic acid, Clarithromycin, Gliclazide, Lercanidipine) in biological matrices is discussed in this work. Relevant literature sources were consulted to understand the different parameters that must be included in method development and validation, to identify what constitutes a good assay method and to know the international regulations pertaining to bioanalytical methodology that determine whether a developed assay method is acceptable or not. Further, literature search is done to collect information on assay methods reported for the selected drugs (discussed in respective chapters). The different aspects of these assay methods *viz.* extraction, instrumentation, total turn-around time and sensitivity, selectivity etc. were assessed. Thus, an objective was set to develop selective, sensitive and rapid UPLC-MS/MS assay methods that have short and simple extraction procedures, consume small amounts of solvent and biological fluid for extraction and rapid compared to existing methods in the literature. Systematic validation as per USFDA guidelines is done for all the developed methods. The parameters investigated include selectivity, sensitivity, cross specificity, carry over effect, linearity, accuracy and precision, absolute and relative recovery, absolute and relative matrix effect, stability in plasma and dilution integrity. The application of these methods for bioequivalence study is conducted with test and reference formulation of the selected drugs on healthy human subjects. The pharmacokinetic parameters investigated include C_{max}, T_{max}, $t_{1/2}$, AUC and K_{el}. Assay reproducibility was demonstrated by incurred samples reanalysis to ensure that the bioanalytical methods developed as per standard guidelines are rugged and reproducible for pharmacokinetic and bioequivalence studies. This study is conducted strictly in accordance with the guidelines laid down by International Conference on Harmonization and USFDA.

1.8 References

[1] World Health Organization, (1969). WHO Expert Committee on Drug Dependence Sixteenth report. (Technical report series. No. 407). Geneva: World Health Organization.

[2] M.R. Siddiqui, Z.A. AlOthman, N. Rahman, Arabian J. Chem.(2013), In Press.

[3] R. Valagaleti, P.K. Burns, M. Gill, Drug Inform. *J.* 37 (2003) 407–438.

[4] C. Han, C. B. Davis, B. Wang, Ed., '*Evaluation of Drug Candidates for Preclinical Development: Pharmacokinetics, Metabolism, Pharmaceutics and Toxicology*', Wiley & Sons Inc., New Jersey (2010).

[5] X. Zhou, R. C. Garner, S. Nicholson, C.J. Kissling, D. Mayers, J. Clin. Pharmacol. 49 (2009) 1408-1416.

[6] N. R. Srinivas, Biomed. Chromatogr.20 (2006) 383-414.

[7] H. Liu, R.H. Liu, D. Lin, H-O. Ho, Rapid Commun. Mass Spectrom. 24 (2009) 75-84.

[8] R. Oertel, K. Richter, T. Gramatte, W. Kirch, J. Chromatogr. A797 (1998) 203-209.

[9] B. K. Matuszewski, M. L. Constanzer, C. M. Chavez-Eng, Anal. Chem. 75 (2003) 3019 3030.

[10]M. Dole, L.L. Mack, R.L. Hines, R.C. Mobley, L.D. Ferguson, M.B. Alice, J. Chem. Phy.49 (1968) 2240-2249.

[11]M. Yamashita, J.B. Fenn, J. Phys. Chem. 88 (1984) 4451-4459.

[12]M. L. Aleksandrov, L. N. Gall, V. N. Krasnov, V. I. Nikolaev, V. A. Pavlenko, V. A. Shkurov, Dokl. Akad. Nauk SSSR277 (1984) 379-383.

[13]S. Zhou, Q. Song, Y. G, W. Naidong, Curr. Pharm. Anal.1 (2005) 3-14.

[14]B. L. Ackermann, M. J. Berna, J.A. Eckstein, L.W. Ott, A. K. Chaudhary,Ann. Review Anal. Chem. 1 (2008) 357-396.

[15]R. Kostiainen, T. Kotiaho, T. Kuuranne, S. Auriola, J. Mass Spectrom. 38 (2003) 357.

[16]D.B. Robb, T.R. Covey, A.P. Bruins, Anal. Chem.72 (2000) 3653-3659.

[17]J.A. Syage, M.D. Evans, Spectroscopy 16 (2001) 15-21.

[18]I.N. Papadoyannis in *'HPLC in Clinical Chemistry'* Chromatographic Science Series, Marcel Dekker Inc., New York, vol. 54 (1990).

[19]Q. Sylvie, W. Christian, P. Alexander, G. Madeleine, S. Caroline, L. Bruno, Ther. Drug Monit. 31 (2009) 695-702.

[20]M.S. Lee, Ed., D.A. Wells in *'Integrated Strategies for Drug Discovery using Mass Spectroscopy'*John Wiley & Sons Inc., New Jersey, (2005) Chapter 17, pp 477-510 (Sample Preparation for Drug Discovery Bioanalysis).

[21] B.K. Matuszewski, M.L. Constanzer, C.M. Chavez-Eng, Anal. Chem. 75 (2003) 3019-3030.

[22] J-M Dethy, B. L. Ackermann, C. Delatour, J.D. Henion, G.A. Schultz, Anal. Chem. 75 (2003) 805-811.

[23] L. Li, F. Liu, X.X. Kong, S. Su, K.A. Li, Chinese Chem. Lett. 13 (2002) 349-350.

[24] P. Juhascik, M. Jenkins, J. Amanda, J. Chromatogr. Sci. 47 (2009) 553-557.

[25] N. Wu, A.M. Clausen, J. Sep. Sci. 30 (2007) 1167-1182.

[26] E. de Hoffmann, V. Stroobant, *Mass Spectromertry: Principles and Applications* (3rd edition), John Wiley and Sons Ltd., Chichester, England (2007).

[27] L.R. Snyder, J.J. Kirkland, J.W. Dolan, *Introduction to Modern Liquid Chromatography* (3rd edition), John Wiley and Sons, Inc., Hoboken, New Jersey, (2010).

[28] K. Robards, P.R. Haddard, P.E. Jackson, *Principles and Practice of Modern Chromatographic Methods*, Academic Press, London, (1994).

[29] S. Ahuja, M.W. Dong, *Handbook of Pharmaceutical Analysis by HPLC* (1st edition), Academic Press, New York (2005).

[30] Y. Wang, F. Ai, S.C. Ng, T.T.Y. Tan, J. Chromatogr. A 1228 (2012) 99-109.

[31] M.A. Niessen, '*Liquid chromatography-mass spectrometry*', 3rd edition, CRC Press, Taylor & Francis, Florida, USA (2006).

[32] H. Lee, J. Liquid Chromatogr. Related Technol. 28 (2005) 1161-1202.

[33] H.T. Karnes, G. Shiu, V.P. Shah, Pharm. Res. 40 (1991) 221-225.

[34] M. Jemal, Biomed. Chromatogr. 14 (2000) 422–429.

[35] P. Bruce, P. Minkkinen, M.L. Riekkola, Mikrochim. Acta.128 (1998) 93-106.

[36] C. Hartmann, J. Smeyers-Verbeke, D.L. Massart, R.D. McDowall, J. Pharm. Biomed. Anal.17 (1998) 193-218.

[37] V.P. Shah, K.K. Midha, S. Dighe, J.I. McGilveray, P.J. Skelly, A. Yacobi, T. Layloff, C.T. Viswanathan, E.C. Cook, R.D. Mcdowall, A.K. Pittman, S. Spector, J. Pharm. Sci. 81 (1992) 309-312.

[38] J. Wieling, G. Hendriks, W.J. Tamminga, J. Hempenius, C.K. Mensink, B. Oosterhuis, J.H.G. Jonkman, J. Chromatogr. A730 (1996) 381-394.

[39] R. Causon, J. Chromatogr. B689 (1997) 175-180.

[40] FDA Guidance for Industry: Bioavailability Studies for Orally Administere Drug-Products-General Considerations, US Department of Health and Human Services, Food and Drug Administration Centre for Drug Evaluation and Research (CDER), (2000).

[41] Guidance for Industry: Bioanalytical Method validation, US Department of Health and Human Services, Food and Drug Administration Centre for Drug Evaluation and Research (CDER), Centre for Veterinary Medicine (CVM), May (2001).

[42] W.A. Ritschel, G.L. Kearns, *Handbook of Basic Pharmacokinetics including Clinical Applications* (6th edition), Apha Publications, New York (2004).

[43] C.T. Vishwanathan, S. Bansal, B. Booth, A.J. DeStefano, M.J. Rose, J. Sailstad, V.P. Shah, J.P. Skelly, P.G. Swann, R. Weiner, AAPS J. 9 (2007) E30- E42.

[44] M.L. Rocci, Jr., V. Devanarayan, D.B. Haughey, P. Jardieu, AAPS J. 9 (2007) E336- E343.

[45] M. Yadav, P.S. Shrivastav, T. de Boer, J. Wieling, P. Singhal, Current Understanding of Bioanalytical Assay Reproducibility – Incurred Sample Reanalysis (ISR), Incurred Sample Stability (ISS) and Incurred Sample Accuracy (ISA). In *Handbook of LC-MS Bioanalysis: Best Practices, Experimental Protocols, and Regulations* (1st edition), W. Li, J. Zhang, F.L.S. Tse (ed.), John Wiley & Sons, Inc. New York (2013) Chapter 5.

[46] M. Yadav, P.S. Shrivastav, Bioanalysis 3 (2011) 1007-1024.

CHAPTER – 2

Experimental - Materials, Instrumentation and Methodology

☞ *CONTENTS*

2.1 MATERIALS, CHEMICALS AND INSTRUMENTATION

2.2 PROTOCOL FOR METHOD VALIDATION AND THE ACCEPTANCE CRITERIA

2.3 BIOEQUIVALENCE STUDY CONDITIONS AND INCURRED SAMPLE REANALYSIS

2.4 REFERENCES

2.1 Materials, chemicals and instrumentation

No.	Instrument	Model	Manufacturer
Table 1. Instrumentation used in the study			
1.	Mass Spectrometer	Quattro Premier XE	Waters, Massachusetts, USA
2.	Liquid Chromatograph	Acquity UPLC	Waters, Massachusetts, USA
3.	Analytical Column	Acquity UPLC BEH C18	Waters, Massachusetts, USA
4.	Autosampler	Acquity UPLC	Waters, Massachusetts, USA
5.	Pump	Acquity UPLC	Waters, Massachusetts, USA
6.	Degasser	Acquity UPLC	Waters, Massachusetts, USA
7.	Column oven	Acquity UPLC	Waters, Massachusetts, USA
8.	Data integration software	MassLynx ^{TM}	Waters, Massachusetts, USA
9.	Analytical Balance	SE-2, BT 224S	Sartorius, Goettingen, Germany
10.	Micropipettes	10-100/100-1000 μL	Eppendorf, Hamburg, Germany
11.	Deep freezer	-20°C, -70°C	Sanyo, Osaka, Japan
12.	Centrifuge machine	Multifuge 3SR	Kendro, Langenselbold, Germany
13.	Refrigerator	2-8°C	Vestfrost, Esbjerg,

			Denmark
14.	Solid phase extractor	Ezypress 48	Orochem, Mumbai, India
15.	Solid phase extraction cartridges	Phenomenex Strata ^{TM}-X (30 mg, 1 cc)	Phenomenex India, Hyderabad, India
16.	Solvent evaporator	Speedovap	Takahe, Mumbai, India
17.	Cyclomixer	Spinix	Tarson, Calcutta, India
18.	Water purifier	MilliQ A10 gradient	Millipore, Bangalore, India
19.	pH meter	Pico+	Labindia, Mumbai, India

This section provides a list of materials, chemicals and solvents, instrumentation, and reference/working standards of selected drugs used during the entire method development, validation and application of the validated method to the study sample analysis. The details are provided in *Table 1, 2 and 3* respectively.

Table 2. Chemicals and solvents used in the study			
Chemicals/Solvents	*Grade*	*Vendor*	*City, Country*
Acetonitrile	HPLC	Mallinckrodt Baker	Estado de Mexico, Mexico
Ammonium formate	Bio ultra	Sigma-Aldrich	St. Louis, MO, USA
Blank human plasma	---	Supratech Micropath	Ahmedabad, India
Formic acid	AR	Sigma-Aldrich	St. Louis, MO, USA
Methanol	HPLC	Mallinckrodt Baker	Estado de Mexico, Mexico
Milli-Q water	Type-I	Millipore	Bangalore, India

| Methyl tert-butyl ether | AR | Merck Specialties Pvt. Ltd. | Mumbai, India |
| n-hexane | AR | Merck Specialties Pvt. Ltd. | Mumbai, India |

Table 3. Working standards used in the study		
Working standards	**Purity**	**Vendor (City, Country)**
Alosetron	99.65%	
Alosetron 13C-d3	99.50%	
Amoxicillin	99.52%	
Amoxicillin-d4	99.36%	
Clavulanic acid	99.67%	Clearsynth Labs (P)
Clarithromycin	99.25%	Ltd.
Clarithromycin 13C-d3	99.58%	(Mumbai, India)
Gliclazide	99.54%	
Gliclazide-d4	99.47%	
Lercanidipine	99.66%	
Lercanidipine-d3	99.53%	

2.2 Protocol for method validation and the acceptance criteria

Method validation plays a significant role in the evaluation and interpretation of bioequivalence study data. The method used for quantitative determination of drugs in biological matrices must generate reproducible and reliable data in order to permit valid interpretation of study data. Selectivity, carry over, matrix effect, accuracy and precision, recovery, dilution integrity and stability under various conditions are the important parameters need to be performed during method validation. The bioanalytical method was thoroughly validated following the USFDA guidelines [1]. The procedures for determining these validation parameters are described below:

System suitability and performance, carryover effect and selectivity

System suitability experiment was performed by injecting six consecutive injections using aqueous standard mixture of analytes and ISs at the start of each batch during method validation as follows,

1. ALO (4.0 ng/mL) and IS (10 ng/mL)
2. AMX/CLV (4500/750 ng/mL) and IS (5000 ng/mL)
3. CLA (600 ng/mL) and IS (100 ng/mL)
4. GLZ (1200 ng/mL) and IS (2.5 μg/mL)
5. LER (8.0 ng/mL) and IS (40 ng/mL)

System performance was studied by injecting one extracted blank (without analytes and ISs) and one extracted LLOQ sample with ISs at the beginning of each analytical batch and before re-injecting any sample during method validation.

Carry over effect of autosampler was checked to verify any carryover of analyte at the start and at the end of each batch. The design of the experiment comprised of the following sequence of injections viz. extracted blank plasma → ULOQ sample → extracted blank plasma → LLOQ sample→ extracted blank plasma.

The selectivity of the method towards endogenous plasma matrix components was assessed in eight different batches of plasma, of which, five were normal K_3EDTA plasma and one each of lipemic, haemolysed and heparinised plasma (except in LER: ten different batches of plasma, of which, six were Na-heparin plasma, two haemolysed and two lipemic).

That was checked by extraction and inspection of the resulting chromatograms for interfering peaks. Interference of commonly used medications by human volunteers was done for paracetamol, chlorpheniramine, caffeine, acetylsalicylic acid and ibuprofen. Their stock solutions (100 µg/mL) were prepared by dissolving requisite amount in methanol. Further, working solutions (1.0 µg/mL) were prepared in the mobile phase and 10 µL was injected for all analytes except 5 µL for AMX and CLV to check for any possible interference at the retention time of analyte and IS.

Linearity, Accuracy and Precision

The linearity of the method was determined by analysis of five linearity curves containing ten non-zero concentrations. The area ratio response for analyte/IS obtained from multiple reaction monitoring was used for regression analysis. Each calibration curve was analyzed individually by using least square weighted ($1/x^2$) linear regression which was finalized during pre-method validation. A correlation coefficient (r^2) value >0.99 was desirable for all the calibration curves. The lowest standard on the calibration curve was accepted as the LLOQ, if the analyte response was at least ten times more than that of drug free (blank) extracted plasma.

For determining the intra-batch accuracy and precision, replicate analysis of plasma samples of analytes were performed on the same day. The run consisted of a calibration curve and six replicates of LLOQ, LQC, MQC-1/2 and HQC samples. The inter-batch accuracy and precision were assessed by analyzing five precision and accuracy batches on three consecutive validation days.

Matrix effect and extraction recovery

Matrix effect is responsible for suppression or enhancement in the measurement of analyte signal due to endogenous or exogenous components present in biological fluids. Matrix effect can directly impact the accuracy, precision, ruggedness and the overall reliability of a validated method. Post-column analyte infusion technique gives a qualitative indication (suppression or enhancement) due to the presence of matrix [2].A standard solution containing analyte and IS was infused post column via a 'T' connector into the mobile phase at 10μL/min employing in-built infusion pump. Aliquots of 10 μL of extracted control plasma were then injected into the column by the autosampler and MRM LC-MS/MS chromatogram was acquired for analyte.

Extraction recovery of analyte and IS from human plasma was evaluated in six replicates by comparing the mean peak area responses of pre-extraction fortified samples to those of post-extraction fortified samples representing 100 % recovery. Absolute matrix effect was assessed by comparing the mean area response of post-extraction fortified samples with mean area of solutions prepared in mobile phase solutions (neat standards) at HQC, MQC-1/2 and LQC levels [3]. IS-normalized matrix factors (MFs, analyte/IS) were calculated to access the variability of the results due to matrix effects. Relative matrix effect was assessed from the precision (% CV) values of the slopes of the calibrations curves prepared from eight different plasma lots/sources, which included one haemolysed and one lipemic plasma. To prove the absence of matrix effect, % CV should be less than 3-4 % for method applicability to support clinical studies [4].

Stability, dilution reliability and method ruggedness

All stability results were evaluated by measuring the area response ratio (analyte/IS) of stability samples against freshly prepared comparison standards with identical concentration. Stock solutions of analytes and ISs were checked for short term stability at room temperature and long term stability at 2-8°C. The solutions were considered stable if the deviation from nominal value was within ±10.0 %. Auto sampler stability, Bench top (at room temperature), wet extract, processed sample stability at room temperature and at refrigerated temperature (4°C) and freeze-thaw stability were performed at LQC and HQC using six replicates at each level. Long term stability of spiked plasma samples stored at -20°C and -70°C was also studied at both these levels. The samples were considered stable if the deviation from the mean calculated concentration of freshly thawed quality control samples was within ±15.0%.

To authenticate the ruggedness of the proposed method, it was performed on two precision and accuracy batches. The first batch was analyzed by two different analysts who were not part of validation of the method while the second batch was studied on two different columns of the same make but different batch. The dilution integrity experiment was performed with an aim to validate the dilution test to be carried out on higher analyte concentrations (above ULOQ), which may be encountered during real subject samples analysis. Dilution integrity experiment was carried out at 2 times the ULOQ concentration. The precision and accuracy for dilution integrity standards at $1/5^{th}$ and $1/10^{th}$ dilution were determined by analyzing the samples against calibration curve standards. The results of the method validation parameters studied for the estimation of drugs,

45

met the acceptance criteria defined in **Table 4**. The formulas used in calculating these parameters are given in **Table 5**.

Sr.No.	Validation Parameter	Acceptance Criteria
	Table 4. Acceptance criteria of validation parameters performed for the drugs	
1	**Calibration curve range**	At least 75% of CSs should fall within ±15% (% bias and %CV)
2	**Selectivity**	Area response at the retention time of analyte and IS in control plasma should be ≤ 20% of LLOQ area & ≤ 5% of IS area respectively
3	**Sensitivity (LLOQ)**	Area response of analyte in LLOQ should be at least five times compared to control plasma area response.
		% CV of concentration should be ≤ 20
		% Nominal concentration should be 80-120
4	**Linearity**	$r^2 \geq 0.9900$
5	**Intra and inter assay accuracy**	% Nominal concentration For LLOQ *:* 80-120 For LQC *:* 85-115 For MQC *:* 85-115 For HQC *:* 85-115 For ULOQ *:* 85-115
6	**Intra and inter assay precision**	% CV For LLOQ *:* ≤ 20 For LQC *:* ≤ 15

		For MQC : ≤ 15
		For HQC : ≤ 15
		For ULOQ : ≤ 15
7	*Recovery of analyte*	% CV across the QC level should be ≤ 20
8	*Recovery of internal standard*	% CV across the QC level should be ≤ 15
9	*Matrix effect*	% Bias / % CV For LQC : ±15 For HQC : ±15
10	*Dilution integrity*	% Nominal concentration For 1/10 of 5×ULOQ : 85-115% For 1/10 of HQC : 85-115 %CV at 1/10 of 5×ULOQ : ≤15 %CV at 1/10 of HQC : ≤15
11	*Stock solution stability of drug, metabolite at room temperature and at refrigerated condition and Intermediate solutions stability at refrigerator temperature (4°± 6°C) {days}*	Mean % change at ULOQ : ±10%
12	*Internal standard stock solution stability at room temperature and at refrigerated condition (4° ± 6°C) {days}*	Mean % change for IS : ± 10%

13	*Stability of drug and its active metabolite in biological matrix (SBM) {h}*	
14	*Refrigerator stability of extracted samples (RSS) {h}*	
15	*Bench top stability of extracted samples (BTS) {h}*	Mean % change:
16	*Freeze and thaw stability of drug and its active metabolite in biological matrices after 6th Freeze thaw cycle at -20°±10°C*	For LQC : ±15 For MQC : ±15 For HQC : ±15
17	*Long term stability of drug and its active metabolite in biological matrices at -20°C {days}*	

Table 5. Common formulas used for calculating the various validation parameters

% CV	$\dfrac{\text{standard deviation} \times 100}{\text{mean}}$
% Accuracy	$\dfrac{\text{calculated conc.} \times 100}{\text{theoretical conc.}}$
% Absolute Recovery[a]	peak area ratio response of samples spiked before extraction × 100 peak area ratio of response of samples spiked in mobile phase (neat sample)
% Relative Recovery	peak area ratio response of samples spiked before extraction × 100

	peak area ratio of response of samples spiked in extracted blank plasma
% Absolute Matrix Effect	$\dfrac{\text{peak area ratio response of samples spiked in extracted blank plasma} \times 100}{\text{peak area ratio of response of samples spiked in mobile phase (neat sample)}}$
% Change	$\dfrac{(\text{stability sample conc.} - \text{comparison sample conc.}) \times 100}{\text{comparison sample conc.}}$

ᵃAlso known as Process Efficiency (PE)

2.3 Bioequivalence study conditions

The purpose of performing bioequivalence studies is to assure that the behavior of the test product is similar to the reference product when given to subjects under identical conditions. Bioequivalence studies need to be performed under strict conditions set by regulatory bodies [1]. Study protocol is prepared and approved by an independent ethical committee. The contents of the study protocol are strictly followed during the studies. It includes back ground information of drug, study design, selection and withdrawal of subjects, treatment of the subjects, assessment of efficacy and safety, ethics, insurance policy, bioanalysis, quality control and assurance, statistics and data handling. All the subjects concerned with the study were informed of the aim and risk involved in the study and a written consent were obtained. The inclusion criteria for volunteer selection (males) was based on the age group between 18 to 45, body mass index (between 18.5 and 24.9 kg/height^2), general physical examination, electro-cardiogram and laboratory tests like hematology, biochemistry and urine examination. All subjects were negative for HIV, HBsAg and HCV tests. The exclusion criteria included allergic responses to any of the drugs studied, volunteers with history of alcoholism, smokers and having a disease which may compromise the haemopoeitic, gastrointestinal, renal, hepatic, cardiovascular, respiratory, central nervous system, diabetes, psychosis or any other body system. The work was approved and subject to review by Institutional Ethics Committee, an independent body comprising of nine members, including lawyers, medical doctors, social workers, pharmacologists and academicians. The procedures followed while dealing with human subjects were based on International Conference on Harmonization, E6 Good Clinical Practice (ICH, E6 GCP) guidelines [5].

The studies design comprised of "An open label, randomized, two period, two treatments, two sequences, balanced, single dose, crossover, comparative evaluation of relative bioavailability of test and reference formulations in healthy Indian human subjects under fasting and/or fed conditions". They were orally administered a single dose of test and reference formulation after recommended wash out period with 240 mL of water. Drinking water was not allowed and supine position was restricted 2h post dose. Dietary and activity restrictions, medications for subjects were followed as per approved study protocol for the subjects participating in the study. Blood samples were collected at pre dose (0.0 h) and after administration of drug at predetermined time intervals. All plasma samples were separated by centrifugation and kept frozen at-20°C till the completion of period and then at -70°C until analysis.

Statistical analysis:-

The plasma concentration-time profiles obtained from the experimental subjects were analyzed by non-compartmental analysis using non-compartmental model using WinNonlin software version 5.2.1 (Pharsight Corporation, Sunnyvale, CA, USA). The C_{max} values and the time to reach maximum plasma concentration (T_{max}) were estimated directly from the observed plasma concentration vs. time data. The area under the plasma concentration–time curve from time 0 to t h (AUC_{0-t}) was calculated using the linear trapezoidal rule. The AUC_{0-inf} was calculated as: $AUC_{0-inf} = AUC_{0-t} + C_t/K_{el}$, where C_t is the last plasma concentration measured and K_{el} is the elimination rate constant; K_{el} was determined using linear regression analysis of the logarithm linear part of the plasma concentration–time curve. The $t_{1/2}$ of NIF was calculated as: $t_{1/2} = \ln2/ K_{el}$. To determine whether the test and reference formulations were pharmacokinetically equivalent, C_{max}, AUC_{0-t}, and AUC_{0-inf} and their ratios (test/reference) using log transformed data were assessed; their means and 90% CIs were analyzed by using SAS® software version 9.1.3 (SAS Institute Inc., Cary, NC, USA). The drug formulations were considered pharmacokinetically equivalent if the difference between the compared parameters was statistically non-significant ($P \geq 0.05$) and the 90% confidence intervals (CI) for these parameters were within 80 to 125 %.

Incurred sample reanalysis:-

An incurred sample reanalysis (assay reproducibility test) was conducted by random selection of 10% of the total number of subject samples analyzed. The selection criteria included samples which were near the C_{max} and the elimination phase in the pharmacokinetic profile of the drug. The results obtained were compared with the data obtained earlier for the same sample using the same procedure. The percent change in the values should not be more than ±20 % [6].

$$\%\text{Change} = \frac{\text{Repeat value} - \text{Initial value}}{\text{Mean of repeat and initial values}} \times 100$$

2.4 References

[1]   Guidance for Industry: Bioanalytical Method validation, US Department of Health and Human Services, Food and Drug Administration Centre for Drug Evaluation and Research (CDER), Centre for Veterinary Medicine (CVM), *May2001*.

[2]   R. King, R. Bonfiglio, C. Fernandez-Metzler, C. Miller-Stein, T.Olah, J. Am. Soc. Mass. Spectrom. 11 (2000) 942–950.

[3]   B.K. Matuszewski, M.L. Constanzer, C.M. Chavez-Eng, Anal. Chem. 75 (2003) 3019-3030.

[4]   B.K. Matuszewski, J. Chromatogr. B 830 (2006) 293-300.

[5]   Guidance for Industry: ICH E6 Good Clinical Practice, U.S. Department of Health and Human Services, Food and Drug Administration, Centre for Drug Evaluation and Research (CDER), Centre for Biologics Evaluation and Research (CBER), April 1996.

[6]   M. Yadav, P.S. Shrivastav, Bioanalysis 3 (2011)1007–1024.

CHAPTER – 3

Analysis of amoxicillin and clavulanic acid by UPLC-MS/MS in human plasma for pharmacokinetic application

☞ *CONTENTS*

3.1   PREAMBLE

3.2   INTRODUCTION

3.3   EXPERIMENTAL AND METHOD OPTIMIZATION

3.3.1 Liquid chromatographic and mass spectrometric conditions

3.3.2 Preparation of calibration standards and quality control samples

3.3.3 Extraction procedure

3.3.4 Validation methodology

3.4   RESULTS AND DISCUSSION

3.4.1 Bioanalytical method development

3.4.2 Method validation results

3.5   BIOEQUIVALENCE STUDY DESIGN AND INCURRED SAMPLE REANALYSIS

3.6   CONCLUSION

3.7   REFERENCE

3.1   Preamble

A bioanalytical method for the simultaneous quantification of amoxicillin and clavulanic acid in human plasma using UPLC-MS/MS has been successfully developed and validated. The analytes and amoxicillin-d4 as internal standard were extracted from 100 μL plasma by solid phase extraction using Phenomenex Strata-X cartridges. Chromatographic separation was done on UPLC BEH C18 (50 mm × 2.1 mm, 1.7 µm) column using acetonitrile: 2.0 mM ammonium formate in water (85:15, *v/v*) as the mobile phase at a flow rate of 0.400 mL/min under isocratic condition. Mass spectrometric detection was by multiple reaction monitoring with an electrospray ionization source in the negative ionization mode. The response of the methodwas linear in the analytical range of 10.0-10000 ($r^2 \geq 0.9997$) and 2.5-2500 ng/mL ($r^2 \geq 0.9993$) for AMX and CLV respectively. Intra- and inter batch accuracy and precision (% CV) were in the range of 96.1-103.2 % and 1.48-5.88 respectively for both the analytes. The mean extraction recovery was 99.6 and 98.4 % for AMX and CLV respectively. Effect of matrix due to endogenous components on the quantitation and the stability of the analytes under different conditions were extensively studied. Dilution reliability and method ruggedness was also evaluated. The method as successfully applied to a bioequivalence study in 12 healthy subjects using 250 mg amoxicillin + 125 mg clavulanic acid fixed dose combination. The assay reproducibility was successfully demonstrated by reanalysis of 63 subject samples.

Keywords: Amoxicillin; clavulanic acid; UPLC-MS/MS; human plasma; solid phase extraction; bioequivalence

3.2 Introduction

Amoxicillin (AMX), a semi-synthetic, acid-stable and an orally absorbed drug. It belongs to a class of antibiotics called the aminopenicillin (β-lactam antibiotic). It is also known to be a 4-hydroxy analog of ampicillin. AMX is extremely effective against various infections caused by wide range of Gram-positive and Gram-negative microorganisms. AMX is prone to degradation by β-lactamase-producing bacteria, which are resistant to β-lactam antibiotics, such as penicillin [1, 2]. Because of this reason, it is often combined with β-lactamase inhibitor, clavulanic acid (CLV) to improve the antibacterial effect and to overcome bacterial resistance. It is a mild antibacterial agent. The name of this resulting drug combination is 'Co-amoxiclav' which reduces susceptibility to β-lactamase resistance, in which the major component is AMX [1]. This potent combination was introduced for the first time in Europe and United States in the year 1981 and 1984 respectively. It is generally used for the treatment of a wide range of bacterial infections, including upper and lower respiratory tract infections and infections of the skin and soft tissue structures [3].

When CLV is co-administered with AMX, there is no any substantial variation of the pharmacokinetics of either drug compared with their separate administration. AMX and CLV are well absorbed by the oral route of administration with peak serum levels appearing within 1-2 h. Protein binding of AMX is very less (17 %) and it is excreted primarily unchanged in the urine with a half-life of about 1 h [3]. Likewise, CLV is extensively metabolized in the liver with a similar half-life and is also slightly protein bound (20-30 %) [4].

Several chromatographic methods, including LC-UV, micellar LC and LC-MS/MS, have been developed for the analysis of AMX as a single analyte [5-7], in presence of its major active metabolites [8] and with other antibiotic drug in binary [9] and ternary combination [10] in different biological matrices. Similarly, number of HPLC-UV [11-14] and LC-ESI-MS/MS [15, 16] methods have been reported for the determination of CLV in various biological fluids. At the same time simultaneous analysis of AMX and CLV is also a subject of several reports [17-26]. Amongst these, three methods have employed HPLC-UV [17-19], while the rest are based on LC-MS/MS technique [20-26]. The simultaneous analysis of AMX and CLV has been done in various biological fluids including human plasma [17, 18, 20-25], human blood [19], and dog plasma [26]. However, several of these methods were either less sensitive [17, 18, 20, 21, 23], had long analysis time [17, 18] or involved large human plasma volume for processing [18, 25]. Additionally, few other methods based on UPLC-MS/MS are presented for the estimation of AMX and CLV along with several other antibiotics [27-29]. A comparative assessment of different methods developed for AMX and CLV in human plasma are presented in *Table 1*.

Based on the existing literature, the main objective of this work was to develop and validate a simple, reliable and rapid method with adequate sensitivity for the simultaneous estimation of AMX and CLV in human plasma by UPLC-MS/MS. The method presents an efficient extraction procedure based on solid phase extraction (SPE) with > 96 % extraction recovery for both the analytes. Systematic evaluation of matrix effect was performed by estimating the IS-normalized matrix factors, relative matrix effect in different plasma sources and through post-column analyte infusion. The proposed method was successfully applied to support a pharmacokinetic study of fixed dose combination of AMX and CLV (250 + 125 mg) tablet formulation in 12 healthy Indian subjects under fasting. The reproducibility in the measurement of study data has been demonstrated by reanalysis of selected incurred samples.

3.3 *Experimental and method optimization*

3.3.1 *Liquid chromatographic and mass spectrometric conditions*

Waters Acquity UPLC BEH C18 (50 mm × 2.1 mm, 1.7 mm) column maintained at 30 °C was used for chromatographic separation of the analytes. Elution of analytes and IS was carried out using a mobile phase consisting of acetonitrile: 2.0 mM ammonium formate in water (85:15, *v/v*), delivered at a flow rate of 0.400 mL/min. The pressure of the system was maintained at 6600 psi. Quantitative determination was performed on Waters Quattro Premier XE (USA) triple quadrupole mass spectrometer equipped with electro ESI in the negative ionization mode.

Table 1. Comparison of salient features of chromatographic methods developed for simultaneous determination of amoxicillin and clavulanic acid in human plasma

Ref.	Technique; linear range (ng/mL)	Extraction procedure; sample volume	Column; mobile phase; flow rate (mL/min); run time (min); retention time (AMX/CLV) in min	Application; matrix effect study; incurred sample reanalysis
[17]	HPLC-UV; AMX: 625-20000 CLV- 312.5-10000	PP; 100 µL	Lichrospher 100 RP8; acetonitrile–phosphate solution-tetramethyl ammonium chloride solution; 1.0; 10.0; 5.48/9.51	Pharmacokinetics study in 3 healthy volunteers with 500 mg AMX and 125 mg CLV; --; --
[18]	HPLC-UV; AMX: 200-12000 CLV: 100-6000	LLE; 500 µL	Chromolith Performance (RP-18e); 0.02 M disodium hydrogen phosphate buffer-methanol (4:96, v/v); 1.3; 10.0; 3.8/5.2	Pharmacokinetics study in 12 healthy volunteers with 500 mg AMX and 125 mg CLV; --; --
[20]	HPLC-ESI-MS; AMX: 125-8000 CLV: 62.5-4000	PP; 200 µL	Zorbax C8; formic acid-water- acetonitrile (2: 1000: 100, v/v/v); 0.4; 3.5; 2.80/2.05	Pharmacokinetic study in 30 healthy volunteers with 250 mg AMX and 125 mg CLV; --; --
[21]	LC-MS/MS; AMX: 500-40000 CLV: 100-6000	LLE; 200 µL	Symmetry C18; acetonitrile -0.1% formic acid (90:10, v/v); 0.5; 3.0; 0.77/0.86	--; --; --
[22]	UPLC-MS; AMX: 10-40000 CLV: 10-10000	PP; 200 µL	Waters Acquity UPLC BEH C18; acetonitrile:water (95: 5, v/v); 0.5; 5.0; 0.92/2.83	Pharmacokinetic study in 24 healthy volunteers with 875 mg AMX and 125 mg CLV; --; --
[23]	LC-MS/MS; AMX: 103-6822 CLV: 46-3026	SPE; 250 µL	Zorbax SB C18; acetonitrile: 2.0 mM ammonium acetate (70: 30, v/v); 0.5; 2.0; 0.86/0.76	--; --; --
[24]	LC-MS/MS; AMX: 5-16000 CLV: 50-2000	LLE; 100 µL	Shim pack XR-ODS; water with 0.2 % formic acid and acetonitrile with 0.2 % formic acid; 0.40; 5.0; 1.22/3.25	Pharmacokinetic study in 12 healthy volunteers with 1000/500/250 mg AMX and 125/62.5/31.25 mg CLV; --; --
[25]	LC-MS/MS; AMX: 50.43-31500.68 CLV: 25.28-6185.18	SPE; 950 µl	HyPURITY advanced C18; acetonitrile -2mM ammonium acetate (80: 20, v/v); 0.8; 1.5; 0.61/0.57	Bioequivalence study in 24 healthy volunteers with 875 mg AMX and 125 mg CLV; --; --
PS	UPLC-MS/MS; AMO: 10-10000 CLV: 2.5-2500	SPE; 100 µl	Waters Acquity UPLC BEH C18; 2mM ammonium formate in water: acetonitrile (15:85, v/v); 0.350; 2.0; 1.57/1.13	Bioequivalence study in 12 healthy volunteers with 250 mg AMX and 125 mg CLV; Yes; Yes (% change within ± 12.0)

AMX: amoxicillin, CLV: clavulanic acid; PP: protein precipitation; LLE: liquid-liquid extraction; SPE: solid phase extraction; PS: present study.

The source dependent parameters maintained for AMX, CLV and IS were, cone gas flow: 110 ± 10 L/h; desolvation gas flow: 800 L/h; capillary voltage: 3.0 kV, source temperature: 100 °C; desolvation temperature: 400 °C; extractor volts: 5.0 V. The pressure of argon used as collision activation dissociation gas was 0.124 Pa. The optimum values for compound dependent parameters like cone voltage and collision energy were set at -26 V and -15 eV for AMX, -32 and -10 eV for CLV, -28 V and -15eV for AMX-d4 respectively. Quadrupole 1 and 3 were maintained at unit mass resolution and the dwell time was set at 100 ms. MassLynx software version 4.1 was used to control all parameters of UPLC and MS.

3.3.2 *Preparation of calibration standards and quality control samples*

The standard stock solutions of AMX (1000 µg/mL) and CLV (500 µg/mL) were prepared by dissolving their requisite amounts in methanol. Further, working solutions were prepared using intermediate solutions of 250.0 µg/mL & 10.0 µg/mL for AMX and 100.0 & 20 µg/mL for CLV in methanol: water (50:50, *v/v*) respectively. Calibration curve standards (CSs) were made at the following concentrations 10.0, 20.0, 50.0, 100.0, 200.0, 500.0, 1000, 2000, 5000, 10000 ng/mL and 2.50, 5.00, 10.0, 25.0, 50.0, 100.0, 250.0, 500.0, 1250, 2500 ng/mL for AMX and CLV respectively. The quality control (QC) samples were prepared at five levels, viz. 8500/2000 ng/mL (HQC, high quality control), 4500/750.0 and 750.0/200.0 ng/mL (MQC-1/2, medium quality control), 30.0/7.50 ng/mL (LQC, low quality control) and 10.0/2.50 ng/mL (LLOQ QC, lower limit of quantification quality control) for AMX/CLV respectively.

Separate stock solutions of the internal standard (100.0 µg/mL for AMX-d4) was prepared by dissolving accurately known amounts of IS in methanol. The working solution was prepared from its stock solutions in methanol: water (50:50, *v/v*) at 5000 ng/mL concentration. Standard stock and working solutions used for spiking were stored at 5 °C, while CSs and QC samples in plasma were kept at -70 °C until use.

3.3.3 Extraction procedure

To an aliquot of 100 µL plasma sample, 25 µL of internal standard was added and vortexed for 10s. Further, 100 µL of 1 % (*v/v*) formic acid in water was added and vortex mixed for another 30s. The samples were centrifuged at 14000 × g for 5 min at 10 °C and loaded on Phenomenex Strata-X (30 mg, 1 cc) cartridge, which was pre-conditioned with 1.0 mL methanol followed by 1.0 mL 1 % (*v/v*) formic acid in water. The samples were washed with 1.0 mL 1 % (*v/v*) formic acid in water followed by 1.0 mL of water. Thereafter, the cartridges were dried for 1 min under nitrogen (1.72 x 10^5 Pa) at 2.4 L/min flow rate. Both the analytes and IS were eluted using 1.0 mL of mobile phase into pre-labeled vials, followed by evaporation to dryness. The dried residue was reconstituted with 250 µL of mobile phase, briefly vortexed for 15 s and 5 µL was used for injection in the chromatographic system using an autosampler.

3.3.4 Validation methodology

The method validation was performed as per the USFDA guidelines [30]. Details of validation procedure and acceptance criteria are given in Chapter 2.

3.4 Results and discussion

3.4.1 Bioanalytical method development

Reported methods for the analysis of AMX and CLV have optimized mass parameters in the positive [20] as well as negative ionization [24] modes as AMX possesses a carboxylic acid and a primary amino group, while CLV has carboxylic acid functionality. Thus, in the present work mass parameters were tuned using ESI in positive as well as negative ionization modes for optimum and consistent response for both the analytes. It was found that AMX gave comparable response in both the ionization modes, while CLV showed much higher response in the negative mode. Thus, to avoid polarity switch both the analytes were analyzed in the negative ionization mode. The Q1 MS full scan mass spectra contained deprotonated precursor $[M-H]^-$ ions at *m/z* 364.1, 197.5 and 368.0 for AMX, CLV and IS respectively. The most abundant and consistent product ions in Q3 MS spectra were observed at *m/z* 223.0, 135.9 and 226.9 for the analytes and IS respectively as shown in ***Figure 1A-C***. The product ion fragment at *m/z* 223.0 and 226.9 for AMX and AMX-d4 corresponded to the possible removal of formic acid and iso-butyl formamide, while the fragment at *m/z* 135.9 for CLV was formed due to elimination of water and carboxylic acid group from the precursor ion. The source dependent and compound dependent parameters were suitably optimized to obtain a consistent and sufficient response for both the analytes. A dwell time of 100 ms gave was adequate date points for the quantitation of the analytes and IS.

The analytes were chromatographically separated on Waters Acquity UPLC BEH C18 (50 mm × 2.1 mm, 1.7 μm) analytical column under isocratic conditions to obtain adequate response and acceptable peak shape. Several existing methods have used acetonitrile together with either formic acid or ammonium acetate as an additive for mobile phase selection. Acetonitrile has been found to be more compatible than methanol with ESI and gives better ionization efficiencies [20]. Thus different combinations of acetonitrile and acidic buffers (ammonium formate/formic acid, ammonium acetate/acetic acid) having different ionic strengths (1-10 mM) in the pH range of 3.0-6.5 were tested. Although acceptable chromatography in terms of baseline separation was observed in allthe systems, the peaks shapes and response was much superior in acetonitrile and ammonium formate solvent system. Further optimization was carried out to finalize their composition by changing the ratio (organic: aqueous) from 50:50 to 90:10 (*v/v*). After several trials, acetonitrile-2.0 mM ammonium formate in water (85: 15, *v/v*), delivered at a flow rate of 0.400 mL/min was optimized to give sharp peak shapes, adequate retention and response. AMX-d4 used as an internal standard worked well in maintaining the ionization efficiency of the analytes and for overall performance of the method.

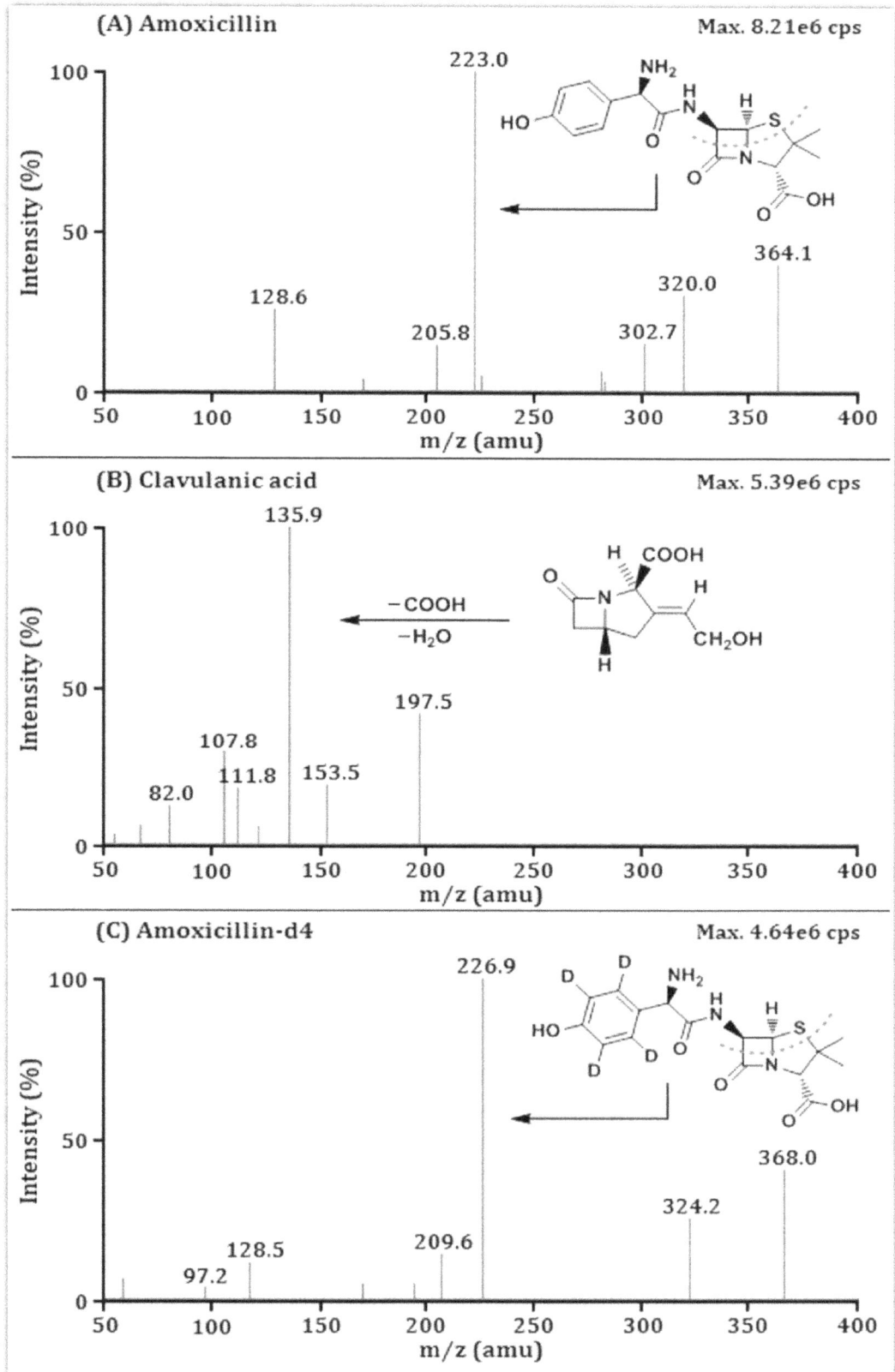

Figure 1.Product ion mass spectra of (A) amoxicillin (m/z 364.1 → 223.0) (B) clavulanic acid (m/z 197.5 → 135.9) (C) amoxicillin-d4, IS (m/z 368.0 → 226.9) in the negative ionization mode, scan range 50-400 amu.

AMX and CLV were baseline resolved within 2.0 min with a retention time was 1.57 and 1.13 min respectively and a resolution factor of 2.09. The reinjection reproducibility (% CV) of retention times for the analytes was ≤ 0.97 % for 60 injections on the same column. The capacity factors, which describe the rate at which the analyte migrates through the column, were 2.83 and 1.41 for AMX and CLV respectively. Representative MRM ion chromatograms in *Figure 2* of (A) blank plasma spiked with IS, (B) AMX and CLV at LLOQ and (C) a real subject sample at C_{max} demonstrates the selectivity of the method to differentiate and quantify the analytes from endogenous components in the plasma matrix or other components in the sample. Further, there was no interference at the retention times of analytes and IS.

All three conventional extraction techniques, protein precipitation (PP) [17, 20, 22], liquid-liquid extraction (LLE) [18, 21, 24] and SPE [23, 25] have been employed for the simultaneous analysis of AMX and CLV from human plasma. Initial trials conducted using different protein precipitants like acetonitrile, methanol and trichloroacetic acid resulted in poor and inconsistent recoveries, especially for CLV (30 to 40 %). Further LLE with common organic diluents, namely dichloromethane, methyl *tert*-butyl ether and *n*-hexane provided somewhat improved recovery compared to PP but matrix interference was also observed. Thus SPE trials were carried out on Phenomenex Strata-X (30 mg, 1 cc) cartridge, which resulted in highly consistent recovery for both the analytes. However, it was found that addition of 100 µL, 1 % (*v/v*) formic acid prior to sample loading on SPE markedly improved the recovery at LLOQ and LQC levels for both the analytes. To establish a method with very low quantitation levels (10.0 ng/mL for AMX and 2.50 ng/mL for CLV), a concentration step was required after elution of samples with 1.0 mL of the mobile phase solution. The extracts obtained were clear with no matrix interference and the recovery was consistent at all QC levels.

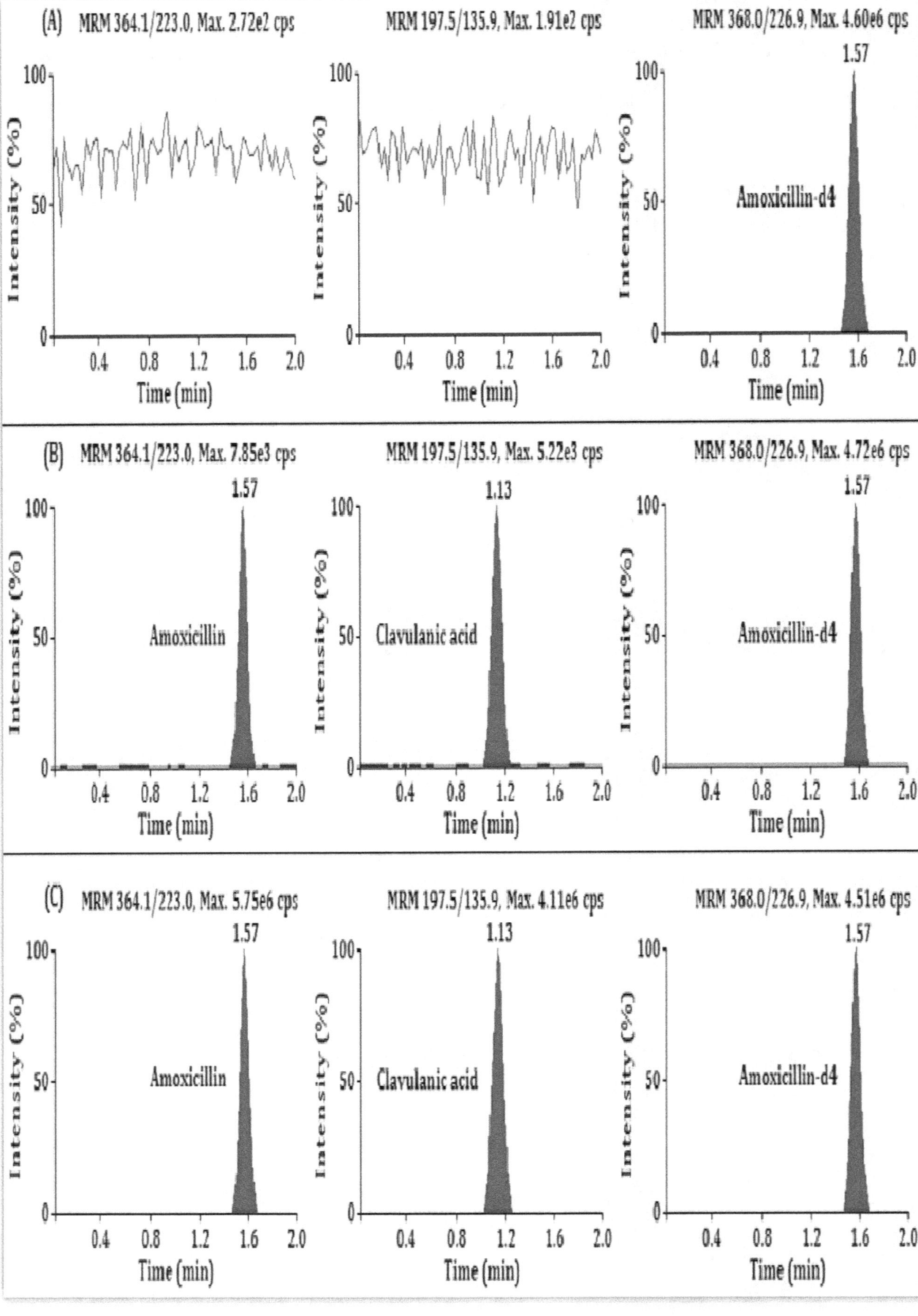

4.2 Method validation results

System suitability, system performance and auto-sampler carryover

The precision (% CV) for system suitability test was observed in the range of 0.37 to 0.84 % for the retention time and 0.65 to 1.45 % for the area response of the analytes and IS. The signal to noise ratio for system performance was $\geq$ 15 for both the analytes. Carry-over evaluation was performed in each analytical run so as to ensure that it does not affect the accuracy and the precision of the proposed method. There was practically negligible carry-over ($\leq$ 0.59 %) in extracted double blank plasma (without analyte and IS) after subsequent injection of highest calibration standard (aqueous and extracted) at the retention time of analytes and IS.

Linearity, lower limit of quantification and accuracy & precision

The five calibration curves were linear over the concentration range of 10.0-10000 and 2.5-2500 ng/mLfor AMX and CLV respectively,with a correlation coefficient (r^2) $\geq$ 0.9993 for both the analytes. The mean linear equations obtained were y = (0.000218 ± 0.000009) x − (0.000064 ± 0.000012) and y = (0.001289 ± 0.000027) x + (0.000048 ± 0.000011) for AMX and CLV respectively. The accuracy and precision (% CV) observed for the calibration curve standards ranged from 98.2 to 101.8 % and 1.18 to 2.98 % for AMX and 97.1 to 103.0 % and 1.12 to 2.95 % for CLV respectively. The lowest concentration (LLOQ) in the standard curve that can be measured with acceptable accuracy and precision was 10.0 and 2.5 ng/mL for AMX and CLV respectively in plasma at a signal-to-noise ratio (S/N) of $\geq$ 15. The LOD values were 2.56 and 0.85 ng/mL for AMX and CLV respectively at S/N $\geq$ 5.The intra-batch and inter-batch precision and accuracy were established from validation runs performed at five QC levels and the results are presented in ***Table 2***.

Table 2. Intra-batch and inter-batch precision and accuracy for amoxicillin and clavulanic acid						
Nominal conc. (ng/mL)	Intra-batch (n = 6; single batch)			Inter-batch (n = 30; 6 from each batch)		
	Mean conc. found (ng/mL)	% CV	Accuracy %	Mean conc. found (ng/mL)	% CV	Accuracy %
Amoxicillin						
8500	8312	2.26	97.8	8670	2.66	102.0
4500	4644	3.05	103.2	4322	3.52	96.1
750.0	767.3	3.97	102.3	739.4	2.79	98.6
30.0	29.65	4.29	98.8	30.33	5.26	101.9
10.0	10.11	5.10	101.1	9.941	4.39	99.4
Clavulanic acid						
2000	2045	1.48	102.3	1966	2.26	98.3
750.0	729.8	3.03	97.3	725.4	2.87	96.7
200.0	202.1	2.94	101.0	198.9	2.80	99.4
7.50	7.37	4.59	98.3	7.718	3.30	102.9
2.50	2.49	5.15	99.6	2.546	5.88	101.8

Extraction recovery and matrix effect

The extraction recovery of analytes from SPE ranged from 96.2-102.3 for AMX and 96.3-101.4 % for CLV. The mean recovery of AMX-d4 was 100.2 %. The presence of endogenous or exogenous components in biological fluids can lead to ion suppression or enhancement in the measurement of analyte signal, giving rise to matrix effect. The chromatogram in *Figure 3a-c* show negligible ion suppression or enhancement at the retention time of analytes and IS. The absolute matrix effect, expressed as matrix factor (MF) was evaluated at four QC levels. The MFs were calculated from the peak area response for the analytes and their IS separately and their ratios were then used to find the IS-normalized MF, which ranged from 0.96-1.04 across four QC levels for both the analytes (*Table 3*). Further, relative matrix effect was assessed in eight different plasma sources. The precision (% CV) in the measurement of slope of calibration lines was 3.92 and 3.13 for AMX and CLV respectively as shown in *Table 4*.

Stability results, method ruggedness and dilution reliability

Stock solutions kept for short-term and long-term stability as well as spiked plasma solutions showed no evidence of degradation under all studied conditions.

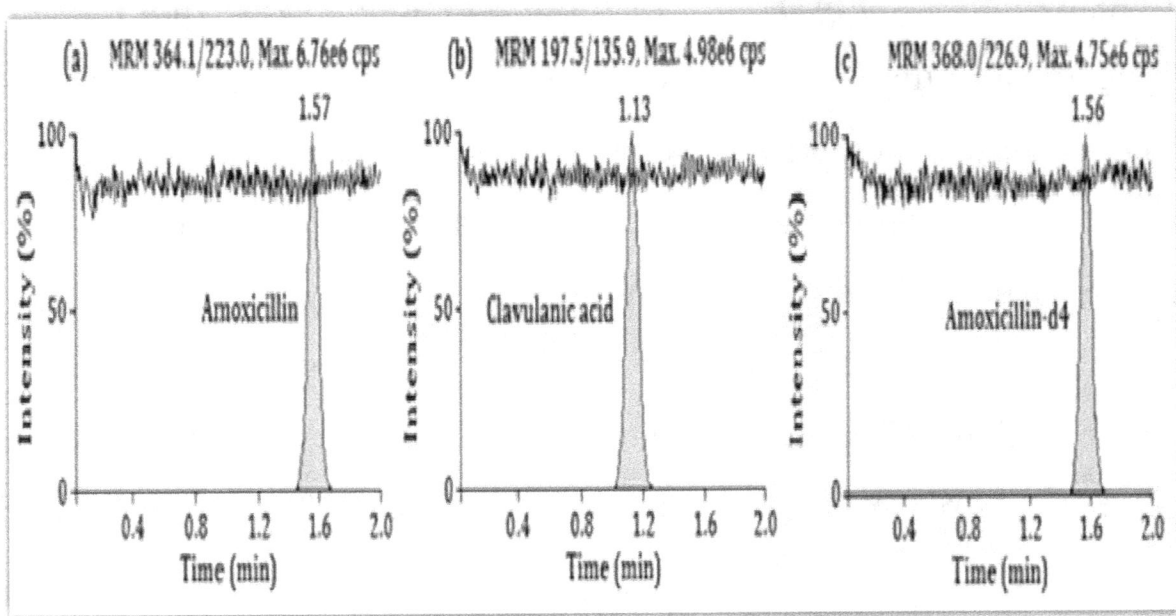

Figure 3.Post column analyte infusion MRM LC-MS/MS chromatograms for (a) amoxicillin, (b) clavulanic acid and (c) amoxicillin-d4.

Table 3. Extraction recovery and matrix effect for amoxicillin and clavulanic acid

QC level	Area response (replicate, $n = 6$)			Extraction recovery, % (B/A)		Matrix factor		
	A	B	C	Analyte	IS	Analyte (A/C)	IS	IS-normalized
Amoxicillin								
LQC	2846	2759	2934	96.2	97.6	0.97	1.01	0.96
MQC-2	70069	70895	70482	101.2	102.4	0.99	0.97	1.02
MQC-1	420064	429814	427951	102.3	100.8	0.98	0.95	1.03
HQC	800594	790365	790169	98.7	99.8	1.01	1.02	0.99
Clavulanic acid								
LQC	1954	1905	1911	97.1	97.6	1.02	0.98	1.04
MQC-2	51315	52056	51833	101.4	102.4	0.99	1.01	0.98
MQC-1	194632	192369	198251	98.8	100.8	0.98	0.97	1.01
HQC	529654	510312	523964	96.3	95.1	1.01	0.98	1.03

A: post- extraction spiking; B: pre- extraction spiking; C: neat samples in mobile phase

No significant degradation of analytes was observed during sample storage and sample processing. The detailed results for stability studies are presented in *Table 5*.

Table 4. Relative matrix effect in eight different lots of human plasma for amoxicillin and clavulanic acid		
Plasma lot	**Slope of calibration curve**	
	Amoxicillin	**Clavulanic acid**
Lot-1	0.000220	0.001250
Lot-2	0.000213	0.001308
Lot-3	0.000221	0.001255
Lot-4	0.000218	0.001311
Lot-5	0.000211	0.001320
Lot-6 (heparinized)	0.000221	0.001241
Lot-7 (haemolysed)	0.000222	0.001273
Lot-8 (lipemic)	0.000197	0.001355
Mean	0.000215	0.001289
±SD	0.000008	0.000040
%CV	3.92	3.13

Table 5. Stability of amoxicillin and clavulanic acid in plasma under different conditions (n = 6)						
	Amoxicillin			**Clavulanic acid**		
Storage Conditions	**Nominal conc. (ng/mL)**	**Mean stability sample (ng/mL)± SD**	**% Change**	**Nominal conc. (ng/mL)**	**Mean stability sample (ng/mL) ± SD**	**% Change**
Bench top stability at 25 °C, 12 h						
	8500	8325 ± 104.9	-2.06	2000	1959 ± 50.19	-2.05
	30.00	31.30 ± 0.51	4.34	7.500	7.625 ± 0.195	1.67
Freeze & thaw stability at -20 °C						
	8500	8655 ± 60.7	1.82	2000	1925 ± 46.40	-3.75
	30.00	29.17 ± 0.41	-2.78	7.500	7.322 ± 0.235	-2.38
Freeze & thaw stability at -70 °C						
	8500	8879 ± 40.2	4.46	2000	2071 ± 49.98	3.55

30.00	29.36 ± 0.28	-2.13	7.500	7.184 ± 0.158	-4.21
Processed sample stability at 25°C , 22 h					
8500	8990 ± 126.3	5.76	2000	2030 ± 56.16	1.50
30.00	30.59 ± 0.86	1.97	7.500	7.921 ± 0.261	5.61
Autosampler stability at 4°C, 72 h					
8500	8123 ± 104.4	-4.44	2000	2042 ± 70.11	2.09
30.00	30.98 ± 0.99	3.28	7.500	7.696 ± 0.192	2.61
Wet extract stability at 2-8°C, 48 h					
8500	8088 ± 102.2	-4.85	2000	1908 ± 36.47	-4.60
30.00	29.75 ± 0.783	-0.82	7.500	7.315 ± 0.125	-2.46
Long term stability at -20 °C, 165 days					
8500	8251 ± 82.50	-2.93	2000	2039 ± 64.42	1.95
30.00	28.59 ± 0.62	-4.70	7.500	7.750 ± 0.120	3.33
Long term stability at -70 °C, 165 days					
8500	8931 ± 151.5	5.07	2000	2105 ± 57.97	5.26
30.00	29.70 ± 0.78	-1.00	7.500	7.225 ± 0.145	-3.67

SD: Standard deviation, n: Number of replicates,

$$\%Change = \frac{Mean\,stability\,samples - Mean\,comparison\,samples}{Mean\,comparison\,samples} \times 100$$

The precision (% CV) and accuracy values for two different columns for method ruggedness ranged from 2.4 to 3.9 % and 95.2 to 104.4 % respectively across five QC levels. For the experiment with different analysts, the results for precision and accuracy were within 1.6 to 2.9 % and 97.3 to 102.4 % respectively at these levels. Fordilution reliability experiment the precision and accuracy values for 1/5th and 1/10th dilution ranged from 1.2-2.5 % and 95.8-103.1 % for both the analytes respectively.

3.5 *Bioequivalence study design and incurred sample reanalysis*

The bioequivalence study was conducted with a single fixed dose of a test (250 mg amoxicillin + 125 mg clavulanic acid tablets from a Generic Company) and a reference (AUGMENTIN®, 250 mg amoxicillin + 125 mg clavulanic acid as potassium salt tablets from GlaxoSmithKline Research Triangle Park, NC 27709, USA) formulation to 12 healthy Indian subjects under fasting conditions. The study was conducted as per International Conference on Harmonization and USFDA guidelines [31]. Blood samples were collected at 0.00 (pre-dose), 0.16, 0.33, 0.50, 0.75, 1.00, 1.25, 1.50, 1.75, 2.00, 2.25, 2.50, 2.75, 3.00, 3.25, 3.50, 4.00, 5.00, 6.00, 8.00, 10.0, 12.0, 14.0, 16.0, 20.0 and 24.0 h after oral administration of the dose for test and reference formulation in labeled K_3EDTA-vacuettes. The plasma concentration *vs.* time profile for AMX and CLV under fasting is shown in ***Figure 4***. The results show that the newly developed method has the required sensitivity to measure plasma concentration of the analytes after oral administration of test and reference formulations. The mean pharmacokinetic parameters evaluated after oral administration of combination tablet are summarized in ***Table 6***.

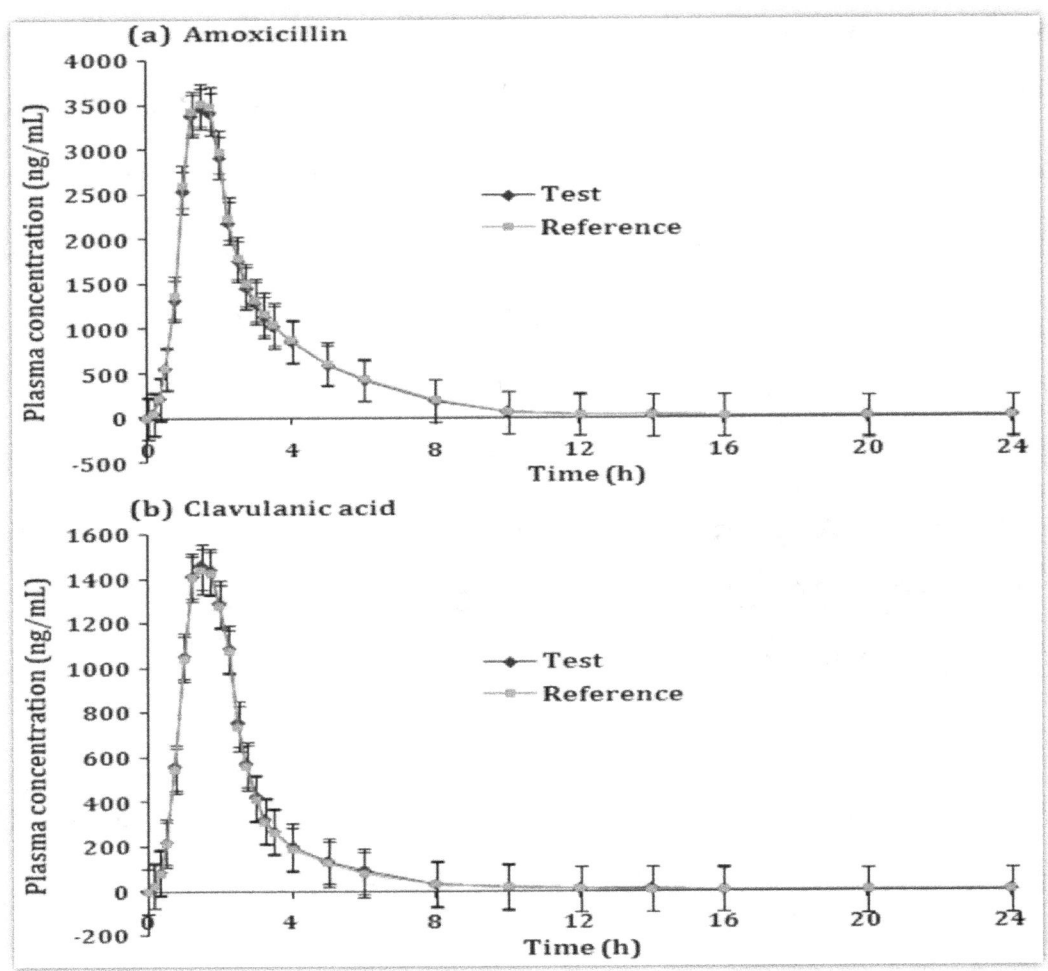

Figure 4. Mean plasma concentration-time profile of amoxicillin and clavulanic acid after oral administration of test (250 mg sumatriptan + 125 mg naproxen sodium tablets from a Generic Company) and a reference (AUGMENTIN®, 250 mg amoxicillin + 125 mg clavulanic acid as a potassium salt from GlaxoSmithKline, USA) formulation to 12 healthy volunteers.

Table 6. Mean pharmacokinetic parameters following oral administration of 250 mg amoxicillin + 125 mg clavulanic acid combination formulation in 12 healthy Indian subjects under fasting

Parameter	Amoxicillin(Mean ±SD)		Clavulanic acid (Mean ±SD)	
	Test	Reference	Test	Reference
C_{max} (ng/mL)	3512.2 ± 360.1	3466.1 ± 329.3	1433.7 ± 154.4	1455.6 ± 165.8

T_{max} (h)	1.57 ± 0.31	1.59 ± 0.23	1.47 ± 0.38	1.52 ± 0.22
$t_{1/2}$ (h)	2.57 ± 0.27	2.50 ± 0.20	2.33 ± 0.17	2.40 ± 0.23
AUC $_{0-24h}$(h.ng/mL)	9619.3 ± 888.6	9377.8 ± 769.9	3244.3 ± 258.1	3350.6 ± 289.7
AUC $_{0-inf}$ (h.ng/mL)	10133.3 ± 951.3	9846.7 ± 868.2	3406.8 ± 333.8	3535.1 ± 361.2
Kel (1/h)	0.270 ± 0.033	0.277 ± 0.051	0.297 ± 0.026	0.279 ± 0.031

The T_{max} and C_{max} values obtained for both the analytes in the present work were comparable with previous report with identical dose strength in 30 healthy subjects [20]. Further, no statistically significant differences were found between the two formulations in any parameter. The ratios of mean log-transformed parameters (C_{max}, AUC_{0-24h} and AUC_{0-inf}) and their 90 % CIs were all within the defined bioequivalence range of 80-125 % (*Table 7*). These observations confirm the bioequivalence of the test sample with the reference product in terms of rate and extent of absorption.

Table 7. Comparison of treatment ratios and 90% CIs of natural log (Ln)-transformed parameters for test and reference formulations in 12 healthy subjects under fasting

Parameter	Ratio (test/ reference), %		90% CI (Lower – Upper)		Power		Intra subject variation, % CV	
	AMX	CLV	AMX	CLV	AMX	CLV	AMX	CLV
C_{max} (ng/mL)	101.3	98.5	97.1-103.8	95.8-102.1	0.9997	0.9998	5.84	4.34
AUC_{0-24h} (h.ng/mL)	102.6	96.8	99.5-104.1	93.6-98.2	0.9993	0.9994	6.24	5.03
AUC_{0-inf} (h.ng/mL)	102.9	96.3	99.9-105.2	93.1-98.6	0.9998	0.9996	7.02	6.32

In order to prove the method reproducibility, 63 study samples were selected which were near the C_{max} and the elimination phase in the pharmacokinetic profile of the drugs. These samples were reanalyzed and the results were compared with the initial study results.

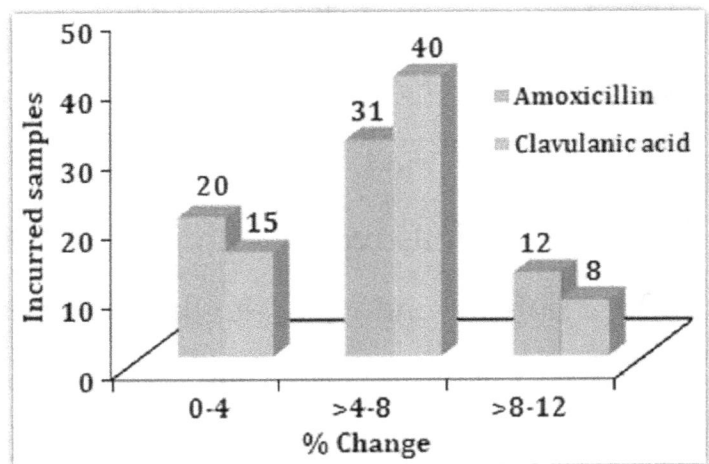

Figure 5. **Graphical representation of results for 63 reanalyzed samples of amoxicillin and clavulanic acid.**

The % change in the results was within ±12 % for both the analytes, which is within the acceptance criterion of ± 20 % [32]. The graphical representation of the results is shown in **Figure 5**.

3.6 Conclusion

The proposed validated method for the estimation of AMX and CLV in human plasma is highly selective, accurate and precise. The method offers significant advantages over those previously reported, in terms of lower sample requirements, sensitivity and analysis time. The efficiency of solid phase extraction and a chromatographic run time of 2.0 min per sample make it an attractive procedure in high-throughput bioanalysis of this antibiotic combination. The linear dynamic range established was adequate to measure the plasma concentration of AMX and CLV in a clinical study involving healthy subjects. In addition, matrix effect and stability of analytes in plasma was extensively studied. Further, incurred sample reanalysis results proved the reproducibility of the proposed method.

3.7 References

[1] J. Kosmidis, D. Willians, J. Andrews, Br. J. Clin. Pract. 26 (1972) 341-346.

[2] H. H. Handsfield, H. Clark, J. F. Wallace, K. K. Holmes, M. Turck, Antimicrob. Agents Chemother. 3 (1973) 262-265.

[3] P. A. Todd, P. Ben Field, Drugs 39 (1990) 264-307.

[4] J. Easton, S. Noble, C. Perry, Drugs 63 (2003) 311-340.

[5] K. M. Matar, Chromatographia 64 (2006) 255-260.

[6] M. R. Alegre, S. C. Broch, J. E. Romero, J. Sep. Sci. 31 (2008) 2813-2819.

[7] S. G. Pingale, M. Badgujar, K. V. Mangaonkar, N. E. Mastorakis, Wseas Transa. Biol. Biomed. 9 (2012) 1-13.

[8] T. Reyns, M. Cherlet, S. De Baere, P. De Backer, S. Croubels, J. Chromatogr. B 861 (2008) 108–116.

[9] M. Szultka, R. Krzeminski, J. Szeliga, M. Jackowski, B. Buszewski, J. Chromatogr. A 1272 (2013) 41–49.

[10] M. A. Khorassani, L. T. Taylor, L. M. Koeth, J. A. Roush, Chromatographia 62 (2005) 459-463.

[11] M. F. Zaater, E. Ghanem, N. Najib, Acta Pharma. Tur. 42 (2000) 50-54.

[12] K. A. Shandi, Euro. J. Sci. Res. 15 (2006) 498-507.

[13] K. M. Matar, E. M. Nazi, Y. M. El-Sayed, M. J. Al-Yamani, S. A. Al-Suwayeh, K. I. Al-Khamis, J. Liq. Chromatogr. Relat. Tech. 28 (2005) 97-107.

[14] H. G. Choi, H. W. Jun, D. D. Kim, H. Sah, B. K. Yoo, C. S. Yong, J. Pharma. Biomed. Anal. 35 (2004) 221-231.

[15] T. Reyns, S. De Baere, S. Croubels, P. De Backer, J. Mass Spectrom. 41 (2006) 1414-1420.

[16] T. Reyns, S. De Boever, S. De Baere, P. De Backer, S. Croubels, Anal. Chim. Acta 597 (2007) 282-289.

[17] G. Hoizey, D. Lamiable, C. Frances, T. Treque, M. Katthieu, J. Denis. H. Millart, J. Pharma. Biomed. Anal. 30 (2002) 661–666.

[18] S. M. Foroutan, A. Zarghi, A. Shafaati, A. Khoddamc, H. Movahed, J. Pharm. Biomed. Anal. 45 (2007) 531–534.

[19] Z. W. Lin, Y. Li, W. J. Song, H. Y. Hu, Y. Zeng, B. H. Xu, J. South Med. Univ. 31 (2011) 1069-1071.

[20] K.H. Yoon, S.Y. Lee, W. Kim, J. S. Park, H. J. Kim, J. Chromatogr. B 813 (2004) 121-127.

[21] A. Ajitha, A. Thenmozhi, D. Sridharan, V. Rajamanickam, M. Palanivelu, Asian J. Pharma. Clin. Res. 3 (2010) 106-109.

[22] B. Ciric, D. Jandrict, V. Kilibarda, J. J. Stosic, V. D. Simict, S. Vucinic, Vojnosanit. Pregl. 67 (2010) 887-892.

[23] K. A. Chaitanya, R. Chelladurai, S. Jeevanantham, R. Vignesh, R. Baskaran, Int. J. Pharm. Pharma. Sci. 4 (2012) 648-652.

[24] J. Zhang, Y. Wang, H. Xie, R. Wang, Z. Jia, X. Men, L. Xu, Q. Zhang, Cell Biochem. Biophys. 65 (2013) 363-372.

[25] A. Gaikwad, S. Gavali, Narendiran, D. Katale, S. Bonde, R. P. Bhadane, J. Pharm. Res. 6 (2013) 804-812.

[26] W. Xi, L. He, C. Guo, Q. Cai, Z. Zeng, Anal. Lett. 45 (2012) 1764-1776.

[27] M. Carlier, V. Stove, J. A. Roberts, E. V. de Velde, J. J. de Waele, A. G. Verstraete, Int. J. Antimicrob. Agents 40 (2012) 416-422.

[28] P. Colin, L. De Bock, H. T'jollyn, K. Boussery, J. V. Bocxlaer, Talanta 103 (2013) 285-293.

[29] R. C. Reyes, R. R. Gonzalez, A. G. Frenchi, M. A. R. Maresca, J. L. M. Vidal, J. Pharm. Biomed. Anal. 89 (2014) 203-212.

[30] Guidance for Industry, Bionanlytical Method Validation, US Department of Health and Human Services, Food and Drug Administration Centre for Drug Evaluation and Research (CDER), Centre for Veterinary Medicine (CVM), May 2001.

[31] Guidance for Industry: ICH E6 Good Clinical Practice, U.S. Department of Health and Human Services, Food and Drug Administration, Centre for Drug Evaluation and Research (CDER), Centre for Biologics Evaluation and Research (CBER), April 1996.

[32] M. Yadav, P.S Shrivastav, Bioanalysis 3 (2011) 1007-1024.

CHAPTER – 4

New improved UPLC-MS/MS method for reliable determination of clarithromycin in human plasma to support a bioequivalence study

☞ *CONTENTS*

4.1   PREAMBLE

4.2   INTRODUCTION

4.3   EXPERIMENTAL AND METHOD OPTIMIZATION

4.3.1 Liquid chromatographic and mass spectrometric conditions

4.3.2 Preparation of calibration standards and quality control samples

4.3.3 Extraction procedure

4.3.4 Validation methodology

4.4   RESULTS AND DISCUSSION

4.4.1 Bioanalytical method development

4.4.2 Method validation results

4.5   BIOEQUIVALENCE STUDY DESIGN AND INCURRED SAMPLE REANALYSIS

4.6   CONCLUSION

4.7   REFERENCES

4.1   Preamble

An improved, highly sensitive UPLC-MS/MS method has been developed for the determination of clarithromycin in human plasma. For sample preparation, liquid-liquid extraction with *n*-hexane: methyl *tert*-butyl ether (20:80, *v/v*) mixture was carried out using clarithromycin 13C-d3 as the internal standard. Acquity UPLC BEH C18 (50 mm × 2.1 mm, 1.7 μm) analytical column was used for chromatography with methanol-5.0 mM ammonium formate, pH 3.0 (78:22, *v/v*) as the mobile phase under isocratic conditions. The analysis time was 1.5 min. Quantitation of analyte was done by tandem mass spectrometer using electrospray ionization in the positive mode. The precursor → product ion transitions monitored for clarithromycin and IS were *m/z* 748.9 → 158.1 and *m/z* 752.8 → 162.0 respectively. The method was validated over a dynamic concentration range of 0.80-1600 ng/mL with correlation coefficient (r^2) ≥ 0.9998. The mean extraction recovery of clarithromycin was 96.2 % across six quality control levels. Intra-batch and inter-batch accuracy and precision (% CV) ranged from 96.8 to 103.5 % and 1.28 to 4.85 % respectively. Stability of clarithromycin in plasma was evaluated under different conditions like bench top, auto sampler, dry and wet extract, freeze-thaw and long term. The present method was successfully applied to a bioequivalence study in 20 healthy subjects who received single oral dose 250 mg clarithromycin tablet formulation. The reproducibility of the method was investigated by reanalysis of 100 incurred samples.

Keywords: ***Clarithromycin, clarithromycin 13C-d3, UPLC-MS/MS, human plasma, sensitive, bioequivalence***

4.2 Introduction

Clarithromycin (6-O-methylerythromycin, CLA) is a semi-synthetic 14-membered macrolide useful for the treatment of a number of bacterial infections. It is active against several Gram-positive (*Staphylococcus aureus, Streptococcus pneumonia, Streptococcus pyogenes*) and Gram-negative (*Haemophilus influenza, Haemophilus parainfluenzae, Moraxella catarrhalis*) bacteria and other microorganisms like *Mycoplasma pneumonia, Chlamydia pneumonia, Mycobacterium avium* and *Mycobacterium intracellulare* [1,2]. The macrolides are a group of antibiotics whose activity is due to the presence of a large macrocyclic lactone ring which has one or more deoxy sugars, usually cladinose and desosamine [3]. CLA is rapidly absorbed from the gastrointestinal tract after oral administration and has an absolute bioavailability of ~ 50%. It is mainly metabolized by cytochrome P450 (CYP) 3A enzymes (CYP3A4 and CYP3A5) to an active 14-hydroxy metabolite. The plasma protein binding of CLA is about 65-70 % and the maximum plasma concentration is attained within 2-4 h [2, 4].

Literature presents several methods to determine CLA either as a single analyte [5-13], or together with its metabolite and other antibiotics [14-20] in different biological matrices like human plasma [5-7, 10-13, 15, 17, 18], rat plasma [8, 16], human serum [9, 14], human urine [11], gastric juice and gastric tissues [16], horse plasma, epithelial lining fluid and broncho-alveolar cells [19] and dried blood spots [20]. In these methods different analytical techniques like capillary electrophoresis [11], HPLC with electrochemical [5, 15, 16], UV [7, 8], fluorescence [9, 14], and mass spectrometric detection [6, 10, 12, 13, 17-20] have been employed. Amongst these only one UPLC-MS/MS based method has been reported with a linear range of 1.0-3000 ng/mL in human plasma [13]. This method presents a chromatographic run time of 2.0 min with 200 μL plasma volume for processing. A comparative summary of salient features of methods developed in human plasma is presented in *Table 1*

In the present work we report an improved UPLC-MS/MS method for determination of CLA in human plasma with respect to the sensitivity, analysis time and plasma sample volume over existing methods. Further, the method employs a deuterated internal standard for better accuracy and precision of the data. The method was applied to a bioequivalence study in 20 healthy subjects using 250 mg clarithromycin tablets.

UPLC-MS/MS METHODS FOR DIFFERENT TYPES OF CATEGORY DRUGS

Sr. No.	Technique; Linear range (ng/ml)	Extraction procedure; internal standard	Sample volume (µL)	Retention time; run time (min)	Application	Ref.
		Table 1. Comparative assessment of chromatographic methods developed for clarithromycin in human plasma				
1	HPLC-electro-chemical; 100-4000	PP with membrane syringe; roxithromycin	120	15.3; 30.0	Pharmacokinetic study in 4 healthy volunteers with a single oral dose of 500 mg clarithromycin tablet	5
2	HPLC-MS/MS; 2.95-20016	LLE with n-hexane-ethyl acetate; roxithromycin	300	1.88; 3.0	Pharmacokinetic study in 9 healthy volunteers with a single oral dose of 500 mg clarithromycin tablet	6
3	HPLC-UV; 31.25-2000	LLE with n-hexane-1-butanol and back extraction with acetic acid; norverapamil	1000	8.37; 10.0	Bioequivalence study with different 250 mg tablet formulations of clarithromycin in 14 healthy volunteers	7
4	LC-MS/MS; 10-5000	PP with acetonitrile; telmisartan	50	1.76; 2.4	Bioequivalence study with 500 mg tablet formulations of clarithromycin in 20 healthy volunteers	10
5	LC-MS/MS; 100-5000	PP with acetonitrile in 96 well plate; roxithromycin	25	1.20; 1.75	Pharmacokinetic study in 6 healthy volunteers with 500 mg tablet formulation of clarithromycin	12
6	UPLC-MS/MS; 1-3000	LLE with diethyl ether; roxithromycin	200	< 1.0; 2.0	Pharmacokinetic study in 18 healthy volunteers with 500 mg tablet formulation of clarithromycin	13
7	HPLC-electroche-mical; 100-10000	PP with methanol and acetonitrile followed by SPE; roxithromycin	500	~13.0; 19.0	Study on eluant composition and column temperature for reproducible electrochemical response	15
8[a]	LC-MS/MS; 36.5-5066.2	PP with acetonitrile; erythromycin	200	1.30; 2.0	Bioequivalence study with 500 mg tablet formulations of clarithromycin in 40 healthy volunteers	17
9[b]	LC-MS/MS; 100-10000	PP with methanol and acetonitrile; cyanoimipramine	100	~2.5; 3.5	--	18
10	UPLC-MS/MS; 0.8-1600	LLE with n-hexane-methyl tert-butyl ether; clarithromycin 13C-d3	100	1.01; 1.50	Bioequivalence study with 250 mg tablet formulations of clarithromycin in 20 healthy volunteers; incurred sample reanalysis with 100 samples	PM

[a]Along with 14-hydroxy clarithromycin. [b]Along with 14-hydroxy clarithromycin, rifampicin and its metabolite; PP: protein precipitation; LLE: liquid-liquid extraction; SPE: solid phase extraction; PM: present method

84

4.3   Experimental and method optimization

4.3.1 Liquid chromatographic and mass spectrometric conditions

A Waters Acquity UPLC system (MA, USA) was used for setting the reversed-phase liquid chromatographic conditions. The analysis of CLA and IS was performed on Acquity UPLC BEH C18 (50 mm × 2.1 mm, 1.7 μm) analytical column maintained at $30°C$ in a column oven. The mobile phase consisted of methanol-5.0 mM ammonium formate, pH 3.0 (78:22, v/v) and was delivered at a flow rate of 0.350 mL/min. The sample manager temperature was maintained at $5°C$ and the pressure of the system was 6300 psi.   Quantitative analysis of CLA and IS was carried out on a Quattro Premier XE (USA) triple quadrupole mass spectrometer from Micro Mass Technologies (MA, USA). It was operated in the positive electro spray ionization mode and the ion transitions monitored for CLA and IS were m/z 748.9/158.1 and m/z 752.8/162.0 respectively. The optimized mass parameters included cone gas flow: 110 L/h, desolvation gas flow: 630 L/h, capillary voltage: 1.4 kV, source temperature: 120 °C, desolvation temperature: 400 °C and extractor voltage: 4.0 V. The optimum values for cone voltage and collision energy were 35 V and 30 eV for CLA and 35 V and 32 eV for IS respectively. MassLynx software version 4.1 was used for data collection and peak integration and to control all parameters of UPLC and mass spectrometer.

4.3.2 Preparation of calibration standards and quality control samples

The standard stock solution of CLA (500.0 μg/mL) was prepared by dissolving requisite amount in methanol. Further, intermediate solutions (100.0 μg/mL and 50.00 μg/mL) for spiking was prepared in methanol:water (60:40, v/v). Calibration standards (CSs) were prepared at

0.80, 1.60, 4.00, 12.00, 36.00, 75.00, 200.0, 400.0, 800.0, 1600 ng/mL and quality control (QC) samples were made at six levels, 1400 ng/mL (HQC, high quality control), 600.0/150.0 ng/mL (MQC-1/2, medium quality control), 2.40 ng/mL (LQC, low quality control) and 0.80 ng/mL (LLOQ QC, lower limit of quantification quality control) for CLA. Stock solution (100.0 µg/mL) of the internal standard was prepared by dissolving 1.0 mg of clarithromycin 13C-d3 in 10.0 mL methanol. Its working solution (100 ng/mL) was prepared by appropriate dilution of the stock solution in methanol:water (60:40, *v/v*). All standard stock and working solutions used for spiking were stored at 5 °C, while CSs and QC samples in plasma were kept at -70 °C until use.

4.3.3 Extraction procedure

Prior to analysis, all frozen subject samples, calibration standards and quality control samples were thawed and allowed to equilibrate at room temperature. To an aliquot of 100 µL of plasma sample, 25 µL of internal standard was added and vortexed for 10s. Further, 2.0 mL of *n*-hexane: methyl *tert*-butyl ether(20:80, *v/v*) solvent mixture was added and vortexed for another 1.0 min. The samples were then centrifuged at 13148 × g for 5 min at 10°C. The organic layer was separated and collected in ria vials, followed by evaporation of the samples at 50 °C under N_2 gas. Finally the samples were reconstituted with 250 µL mL of mobile phase solution, briefly vortexed and 10 µL was used for injection in the chromatographic system.

4.3.4 Validation methodology

The method validation was performed as per the USFDA guidelines [21]. Details of validation procedure and acceptance criteria are given in *Chapter-2*.

4.4   Results and discussion

4.4.1 Bioanalytical method development

The present work was intended to improve upon the existing UPLC-MS/MS method [13] with respect to the sensitivity, analysis time and plasma sample volume for reliable determination of CLA in human plasma. Mass spectrometric conditions like collision energy, cone voltage and capillary voltage were suitably optimized to obtain maximum sensitivity for CLA and CLA 13C-d3 (IS). Unlike other methods which employed a general IS, a deuterated IS was used for better accuracy and precision of the obtained data. MS/MS analysis was done using electrospray ionization in the positive mode to achieve high selectivity and a good linearity in regression curves. The full scan mass spectra for CLA and IS predominantly contained precursor $[M+H]^+$ ions at *m/z* 748.9 and 752.8 respectively. The most abundant and consistent product ions in Q3 MS spectra for CLA and IS were observed at *m/z* 158.1 and 162.0 respectively *(Figure 1)*.

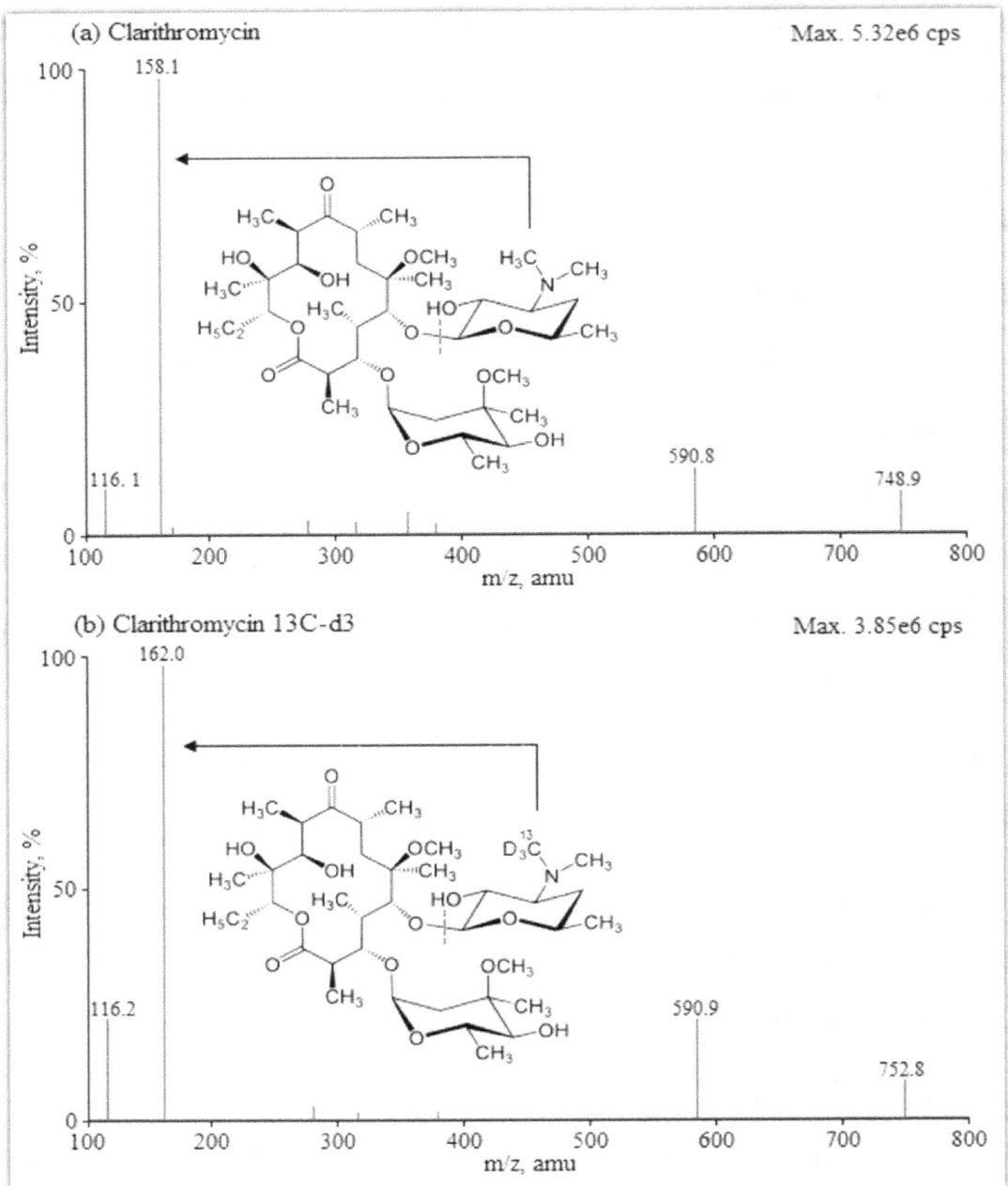

*Figure 1.**Product ion mass spectra of (a) clarithromycin (m/z 748.9 → 158.1) and (b) internal standard, clarithromycin 13C-d3 (m/z 752.8 → 162.0) in the scan range 100-800 amu and in the positive ionization mode.***

These product ions were formed by the breaking up of glycosidic bond to obtain fragments containing tertiary amino group from the protonated precursor ions. A dwell time of 100 ms for both the compounds was sufficient to obtain sufficient to obtain at least 23 data points for quantification. The current UPLC-MS/MS method was more sensitive (0.8 ng/mL) compared to all reported methods for CLA.

Sample preparation is crucial for reliable quantitation of drugs in biological samples. Several reported methodologies have adopted either protein precipitation (PP) [5, 10, 12, 15] or liquid-liquid extraction (LLE) [6, 7, 13] for sample clean-up. Thus, PP was initially tested with acetonitrile and methanol as discussed earlier [10, 15]. However, due to significant matrix effect the recovery of CLA was very poor (40-50 %), especially at LLOQ and LQC levels. Thus, LLE was tried in different solvents systems like diethyl ether, methyl *tert*-butyl ether, n-hexane and ethyl acetate as the CLA has high logP value of 3.2. Further, in our effort to have an efficient solvent extraction protocol for CLA and IS their combinations were also tested. After several trials it was found that *n*-hexane:methyl*tert*-butyl ether(20:80, *v/v*) was the best extraction solvent system for quantitative and precise recovery of the analyte and IS from plasma samples. Only 100 µL plasma sample was used for processing unlike the other UPLC-MS/MS method [13] which employed 200 µL sample.

The nature of mobile phase and its composition, buffer pH and flow rate was rigorously optimized on UPLC BEH C18 (50 mm × 2.1 mm, 1.7 µm) analytical column so as to have adequate detector response and acceptable peak shape. Initially, organic diluents like acetonitrile/methanol along with acidic buffers (ammonium formate-formic acid/ammonium acetate-acetic acid) in the pH range of 3.0-5.5 were tested. Additionally, the concentration of ammonium formate and ammonium acetate was also tested, including different flow rates (0.250, 0.300, 0.350 and 0.400

mL/min) for adequate response. It was found that methanol and ammonium formate-formic acid buffer provided superior response compared to acetonitrile and ammonium acetate-acetic acid buffer.

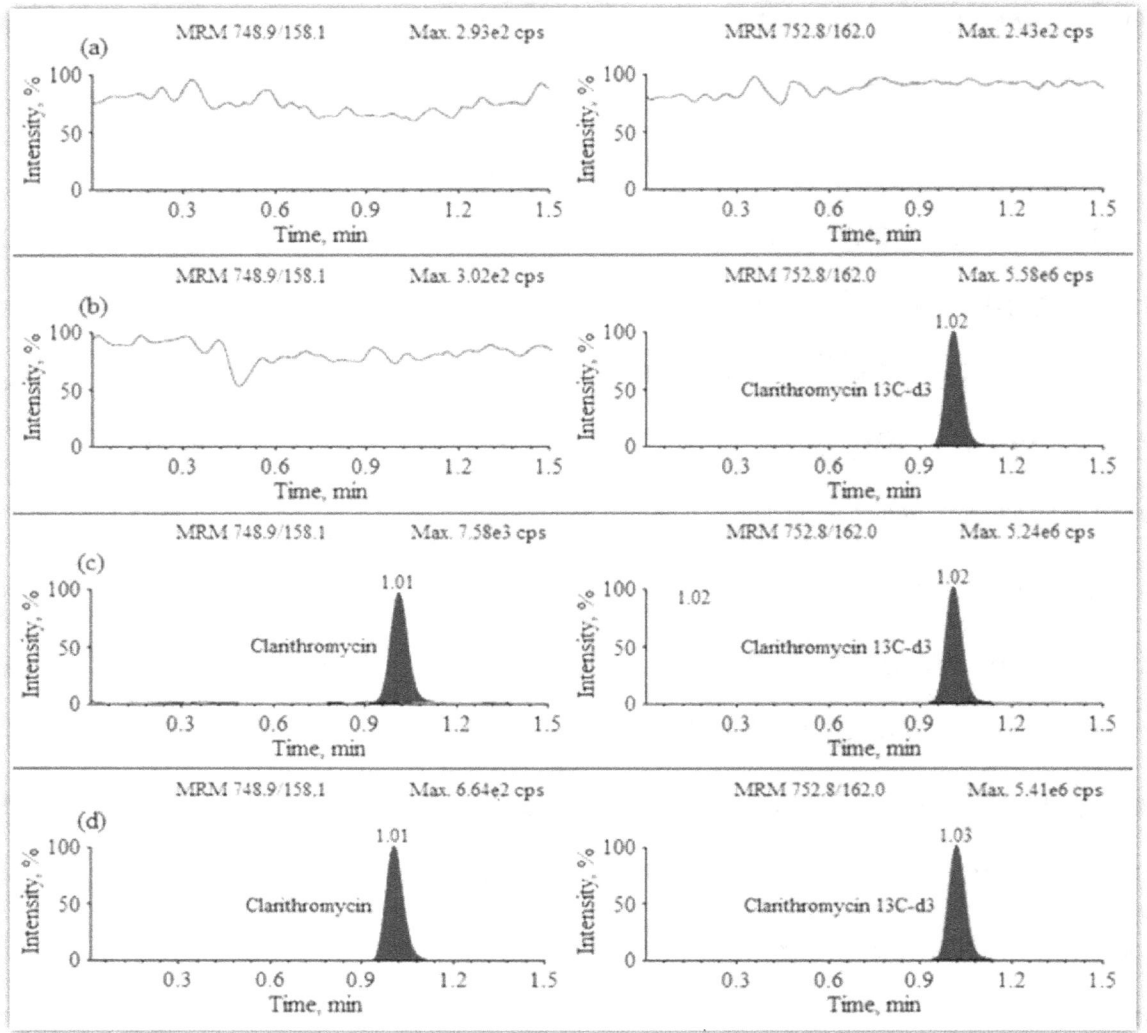

Figure 2. **Representative MRM ion-chromatograms of (a) double blank plasma without analyte and IS, (b) blank plasma with working solution of clarithromycin 13C-d3 (m/z 752.8 → 162.0), (c) clarithromycin (m/z 748.9 → 158.1) at LLOQ and IS, (d) clarithromycin in real subject sample at Cmax and IS after administration of 250 mg dose of clarithromycin.**

Thereafter, their ratio was adjusted for adequate retention, response and better peak shape for CLA and IS. The best chromatographic conditions were obtained with methanol and 5.0 mM ammonium formate, pH 3.0 adjusted with 0.1% formic acid (78:22, *v/v*) as the mobile phase at a flow rate of 0.350 mL/min. These conditions afforded a run time of 1.5 min with retention times of 1.01 and 1.02 min for CLA and IS respectively. This analysis time was shorter compared to the existing UPLC-MS/MS method.The reinjection reproducibility (% CV) of retention times for CLA was ≤ 0.91 % for one entire batch on the same column. Clarithromycin 13C-d3, used as IS in the present work gave excellent results for accuracy and precision at each QC level. The multiple reaction monitoring (MRM) chromatograms illustrated in *Figure 2* of extracted blank plasma (double blank), blank plasma fortified with IS, CLA at 0.8 ng/mL and IS and a subject sample at C_{max} demonstrates the overall performance and selectivity of the method.

4.4.2 Method validation results

The precision (% CV) of system suitability test was observed in the range of 0.26 to 0.51 % for the retention time and 0.84 to 1.28 % for the area response for CLA and IS, while the signal to noise ratio for system performance was ≥ 20. Carry-over evaluation was performed in each analytical run so as to ensure that it does not affect the accuracy and the precision of the proposed method. The column and auto-sampler carry-over evaluation showed negligible carry over in blank plasma (≤ 0.38 % of LLOQ sample) after subsequent injection of ULOQ sample (chromatograms not shown) at the retention time of CLA and IS.

There was adequate linear correlation between the CLA concentration and the response over the concentration range of 0.80-1600 ng/mLwith correlation coefficient (r^2) ≥ 0.9998.The mean linear equation for the calibration curves was y = (0.001632 ± 0.000004)x - (0.000013 ±

0.000011). The accuracy and precision (%CV) observed for the calibration curve standards ranged from 97.7 to 101.7 % and 1.03 to 2.90 % respectively *(Table 2)*. The analytical method was shown to be selective based on absence of any analytical signals at the retention time of CLA and IS in ten different batches of blank plasma. The intra-batch and inter-batch precision and accuracy results were within the stipulated range of ±15 % of the nominal concentration and □ 15 % CV of the mean values as shown in *Table 3*.

Matrix effect can be attributed to some undesirable effects that originate from a biological matrix. These components may result in ion suppression/enhancement, decrease/increase in sensitivity of analyte over a period of time, increased baseline, imprecision of data, drift in retention time and distortion or tailing of a chromatographic output. It is suggested that evaluation of matrix factors (MFs) can help to assess the matrix effect. Further, matrix effect needs to be checked in lipemic and haemolysed plasma samples in addition to normal K_3EDTA plasma. MFs can be determined from the peak area response for the analyte and IS separately, while the ratio of the two factors yields IS-normalized MF. The IS-normalized MFs using stable-isotope labelled IS should be close to unity because of the similarity in the chemical properties and elution times for the analyte and IS. The extraction recovery and matrix factors for CLA are presented in *Table 4*.

The mean extraction recovery for CLA and IS was 96.2 and 96.7 % respectively. The relative matrix effect in eight different plasma sources was evaluated by calculation of precision (% CV) of the slopes of calibration lines *(Table 5)*. Further, qualitative assessment of matrix effect through post-column infusion experiment showed no ion suppression or enhancement at the retention time of CLA and IS in the chromatograms *(Figure 3)*.

Table 2. Summary of calibration curve with back calculated concentration for clarithromycin

	STD-1	STD-2	STD-3	STD-4	STD-5	STD-6	STD-7	STD-8	STD-9	STD-10	Regression Parameters		
	Nominal concentration (ng/mL)												
ID No.	0.800	1.600	4.090	12.00	36.00	75.00	200.0	400.0	800.0	1600	Slope	Intercept	r^2
1	0.815	1.546	4.098	12.36	36.65	75.91	202.3	403.9	808.1	1629	0.00163	- 0.00000456	0.9999
2	0.784	1.648	3.911	11.81	35.69	74.08	197.9	397.3	794.3	1575	0.00163	-0.0000209	0.9998
3	0.809	1.565	4.108	11.72	36.41	76.21	203.8	404.5	809.5	1635	0.00163	-0.0000026	0.9999
4	0.789	1.639	3.901	12.27	35.72	73.89	198.5	395.4	791.7	1561	0.00164	-0.0000273	0.9998
5	0.813	1.573	4.115	11.78	36.28	74.11	203.1	398.1	807.2	1618	0.00163	-0.0000090	0.9999
Mean	0.802	1.594	4.027	11.99	36.15	74.84	201.1	399.8	802.2	1603.6	0.001632	-0.000013	0.9999
S.D.	0.0144	0.0462	0.1103	0.3019	0.4275	1.1219	2.7262	4.1046	8.4521	33.4335	0.000004	0.000011	0.000054
%CV	1.80	2.90	2.74	2.52	1.18	1.50	1.36	1.03	1.05	2.08			
% Nominal	99.8	100.5	99.1	101.2	101.7	101.1	98.0	100.8	97.7	100.9			

CV: coefficient of variance; S.D.: standard deviation; r^2: correlation coefficient

Table 3. Intra-batch and inter-batch precision and accuracy data for clarithromycin

Nominal concentration (ng/mL)	Intra-batch (n = 6; single batch)			Inter-batch (n = 30; 6 from each batch)		
	Mean conc. found (ng/mL)	% CV	% Accuracy	Mean conc. found (ng/mL)	% CV	% Accuracy
HQC (1400)	1405.6	2.80	100.4	1393.0	1.28	99.5
MQC-1 (600.0)	613.2	3.23	102.2	585.6	3.59	97.6
MQC-2 (150.0)	147.5	2.91	98.3	154.3	2.84	102.9
LQC (2.40)	2.484	2.89	103.5	2.374	2.94	98.9
LLOQ QC (0.80)	0.774	4.85	96.8	0.825	3.68	103.2

Table 4. Extraction recovery and matrix factor for clarithromycin

QC level	Area response (replicates, n = 6)			% Extraction recovery, (B/A)		Matrix factor		
	A	B	C	Analyte	IS	Analyte (A/C)	IS	IS-normalized
LQC	2309	2205	2127	95.5	97.0	1.085	1.077	1.007
MQC	14812	14212	13618	96.0	96.5	1.087	1.08	1.004

-2	4	7	9				2	
MQC	58687	56655	54973	96.5	97.4	1.067	1.07	0.996
-1	6	4	8				1	
HQC	13715	13287	12953	96.9	95.9	1.058	1.06	0.991
	49	53	64				7	

A: post-extraction spiking; B: pre-extraction spiking; C: neat samples in mobile phase

Table 5. Relative matrix effect in eight different lots of human plasma for clarithromycin	
Plasma lot	**Slope**
Lot-1 (K_3EDTA)	0.001651
Lot-2 (K_3EDTA)	0.001615
Lot-3 (K_3EDTA)	0.001648
Lot-4 (K_3EDTA)	0.001618
Lot-5 (K_3EDTA)	0.001653
Lot-6 (heparinized)	0.001608
Lot-7 (haemolysed)	0.001658
Lot-8 (lipemic)	0.001606
Mean	0.001632
± Standard deviation	0.000022
% Coefficient of variation	1.36

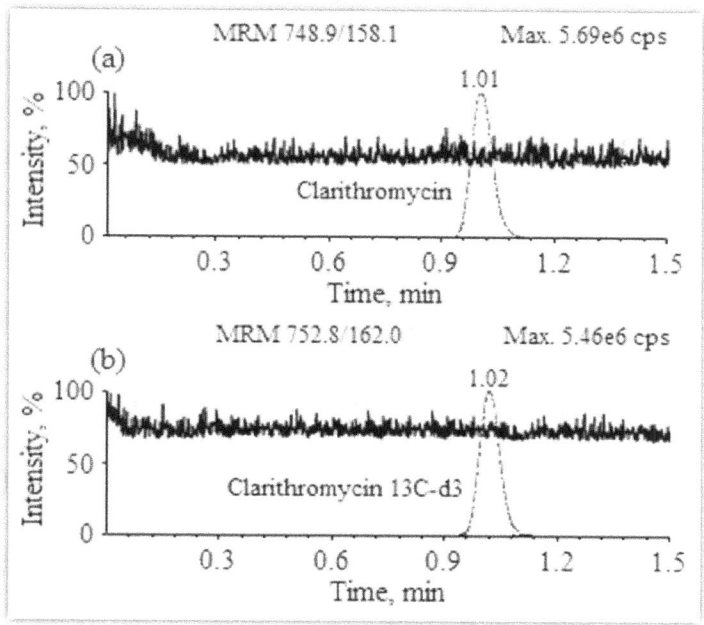

*Figure 3.**Injection of extracted blank human plasma during post column infusion of (a) clarithromycin at 800 ng/mL and (b) clarithromycin 13C-d3 at100 ng/mL.*

The stability of analyte and IS in human plasma and stock solutions was examined under different storage conditions. Stock solutions for short term stability of CLA and IS were stable at room temperature up to 24 h and between 2-8°C for a minimum period of 30 days. CLA in control human plasma (bench top) at room temperature was stable for at least 20 h at 25°C and for minimum of five freeze and thaw cycles. Autosampler (processed sample) stability of the spiked quality control samples was determined up to 52 h. Similarly, dry and wet extract stability of the samples was ascertained up to 24 h and 48 h respectively. Long term stability of the spiked quality control samples remained unchanged up to 176 days. The % change values for different stability experiments at LQC and HQC levels in plasma are shown in *Table 6*.

Table 6. Stability results of clarithromycin in plasma under various conditions (n = 6)				
Storage conditions	QC Level	Nominal conc.	Mean stability	Change (%)

		(ng/mL)	sample (ng/mL) ± SD	
Bench top stability at 25 °C, 20 h	HQC	1400	1404.80 ± 37.07	0.34
	LQC	2.40	2.407 ± 0.070	0.31
Freeze-thaw stability at -20 °C	HQC	1400	1410.60 ± 38.93	0.76
	LQC	2.40	2.409 ± 0.037	0.39
Freeze-thaw stability at -70 °C	HQC	1400	1391.80 ± 37.85	-0.59
	LQC	2.40	2.390 ± 0.046	-0.42
Autosampler stability at 4°C, 52 h	HQC	1400	1397.60 ± 39.68	-0.39
	LQC	2.40	2.382 ± 0.057	-0.72
Dry extract stability at 2-8°C, 24 h	HQC	1400	1408.80 ± 37.79	0.63
	LQC	2.40	2.386 ± 0.042	-0.57
Wet extract stability at 2-8°C, 48 h	HQC	1400	1389.00 ± 40.70	-0.79
	LQC	2.40	2.411 ± 0.039	0.49
Long term stability at -20 °C, 176 days	HQC	1400	1392.60 ± 32.78	-0.53
	LQC	2.40	2.404 ± 0.031	0.18
Long term stability at -70 °C, 176 days	HQC	1400	1406.60 ± 24.26	0.47
	LQC	2.40	2.394 ± 0.026	-0.22

The precision (% CV) and accuracy values for two different columns ranged from 2.12 to 3.34 % and 96.1 to 103.9 % respectively at all five quality control levels. For the experiment with different analysts, the results for precision and accuracy were within 1.64-3.41 % and 97.6 to

100.1 % respectively at these levels. The dilution integrity experiment was performed with an aim to validate the dilution test to be carried out on higher analyte concentration above the upper limit of quantification (ULOQ), which may be encountered during real subject sample analysis.The precision and accuracy values for 1/2th and 1/10th dilution ranged from 2.97 to 4.57 % and 96.9-101.3 % respectively for CLA.

4.5 Bioequivalence study design and incurred sample reanalysis

The validated method was successfully applied for the assessment of bioequivalence of a test (250 mg clarithromycin tablets from a Generic Indian Company, India) with a reference (BIAXIN® FILMTAB®, 250 mg clarithromycin tablets from Abbott Laboratories, North Chicago, USA) formulation in 20 healthy Indian males under fasting. The study was performed according to International Conference on Harmonization and USFDA guidelines [22]. Blood samples were collected at 0.00 (pre-dose), 0.33, 0.67, 1.00, 1.33, 1.67, 2.00, 2.33, 2.67, 3.00, 3.33, 3.67, 4.0, 4.5, 5.0, 5.5, 6.0, 8.0, 10.0, 12.0, 16.0, 24.0, 36.0, 48.0 and 72.0 h in K_3-EDTA vacutainers. **Figure 4** shows the plasma concentration vs. time profile of clarithromycin after oral administration of 250 mg dose under fasting.

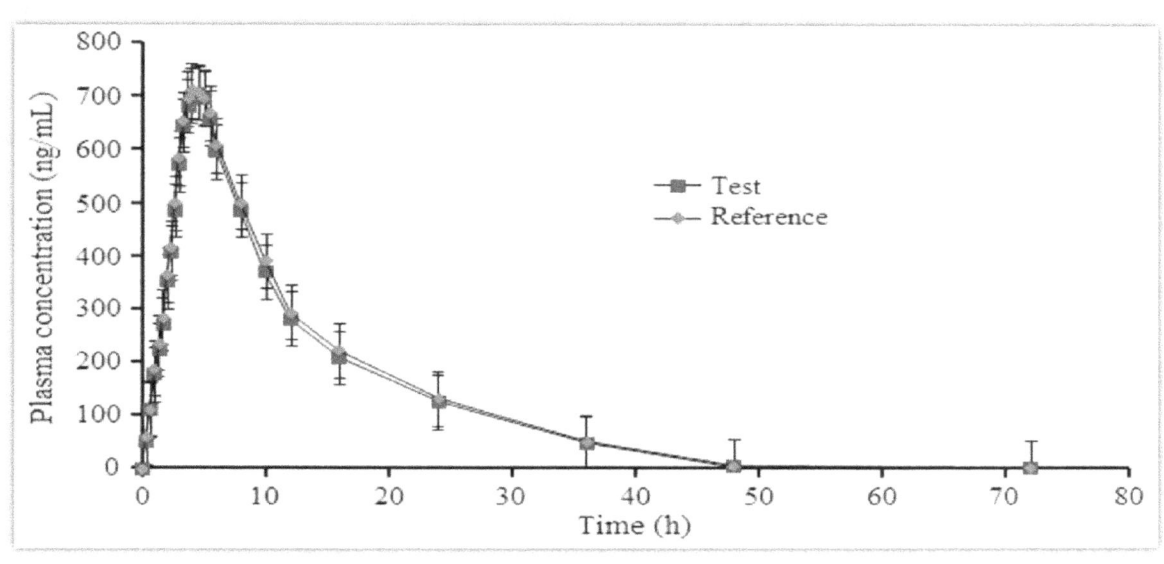

Figure 4. Mean plasma concentration-time profile of clarithromycin after oral administration of test (250 mg tablets from a Generic Indian Company, India) with a reference (BIAXIN® FILMTAB®, 250 mg tablets from Abbott Laboratories, USA) formulation to 20 healthy subjects.

Approximately 900 samples including the calibration and QC samples with volunteer samples were run and analyzed during a period of 5 days and the precision and accuracy for calibration and QC samples were well within the acceptable limits. The important pharmacokinetic parameters, maximum plasma concentration (C_{max}), area under the plasma concentration-time curve from 0 to 72 h (AUC_{0-72}), area under the plasma concentration-time curve from zero hour to infinity (AUC_{0-inf}), time point of maximum plasma concentration (T_{max}), half-life of drug elimination during the terminal phase ($t_{1/2}$) and elimination rate constant (K_{el}) were calculated for the test and reference formulations and are presented in **Table 7**. The T_{max} and $t_{1/2}$ values were in agreement with a reported study in healthy Indian subjects [17].

Further, the 90 % confidence interval (CI) of individual ratio geometric mean for test/reference was within 80-125 % for AUC_{0-72}, AUC_{0-inf} and C_{max} *(Table 8)*. These observations confirm the bioequivalence of the test sample with the reference product in terms of rate and extent of absorption. The incurred sample reanalysis (ISR) results showed % change values to be within ±13 % for CLA for selected samples, which is within the acceptance criterion of ± 20 % [23]. This confirms the reproducibility of the proposed method.

Table 7. Mean pharmacokinetic parameters after oral administration of 250 mg clarithromycin tablet formulation in 20 healthy Indian subjects under fasting.

Parameter	Test (Mean ±SD)	Reference (Mean ±SD)
C_{max} (ng/mL)	700.15 ± 223.27	711.09 ± 214.82
$AUC_{0\text{-}72\,h}$ (h. ng/mL)	8844.2 ± 345.54	9077.3 ± 356.18
$AUC_{0\text{-}inf}$ (h. ng/mL)	8915.5 ± 399.35	9152.2 ± 389.79
T_{max} (h)	4.07 ± 1.02	4.01 ± 1.09
$t_{1/2}$ (h)	9.01 ± 2.15	9.39 ± 2.78
Kel (1/h)	0.172 ± 0.005	0.170 ± 0.007

Table 8. Treatment ratios and 90% CIs of natural log (Ln)-transformed parameters of test and reference tablet formulations of clarithromycin in 20 healthy subjects under fasting.

Parameter	Ratio (test/reference) , %	90% CI (Lower – Upper)	Power	Intra subject variation, % CV
C_{max}	98.5	95.1-102.8	0.9999	3.49
$AUC_{0\text{-}72\,h}$	97.4	95.2-100.5	0.9995	5.72
$AUC_{0\text{-}inf}$	97.2	94.7-102.2	0.9998	4.25

4.6 Conclusion

A highly sensitive and rapid UPLC-MS/MS method for the determination of CLA in human plasma has been developed and fully validated. The method was shown to be selective and free from matrix interference as evident from the results of post-column infusion, IS-normalized matrix factors and relative matrix effect in different plasma lots. The validation results indicate good linearity, accuracy and precision, recovery and stability of CLA in human plasma. With twofold dilution reliability, it is possible to extend the upper limit of quantitation to 3200 ng/mL. The developed method was successfully applied to analyze CLA concentration in pharmacokinetic study. Further, an incurred sample reanalysis of 100 selected samples confirms the reproducibility of the proposed method, which is not reported in any previous method.

4.7 *References*

[1] D. H. Peters, S. P. Clissold, Drugs 44 (1992) 117-164.

[2] BIAXIN® FILMTAB® (clarithromycin tablets), Prescribing Information, June 2012. Manufacture by Abbott Laboratories, North Chicago, IL 60064, USA. http://www.drugs.com/ pro/biaxin.html

[3] M. Ogrendik, Int. J. Gen. Med. 7 (2014) 43-47.

[4] A. D. Rodrigues, E. M. Roberts, D. J. Mulford, Y. Yao, D. Ouellet, Drug Metab. Dispos. 25 (1997) 623-630.

[5] S. J. Choi, S,B. Kim, H. Y. Lee, D. H. Na, Y. S. Yoon, S. S. Lee, J. H. Kim, K. C. Lee, H. S. Lee, Talanta 54 (2001) 377-382.

[6] G. F. V. Rooyen, M. J. Smit, A. D. D. Jager, H. K. L. Hundt, K. J. Swart, A. F. Hundt, J. Chromatogr. B 768 (2002) 223-229.

[7] H. Amini, A. Ahmadiani, J. Chromatogr. B 817 (2005) 193-197.

[8] W. Li, H. Jia, K. Zhao, Talanta 71 (2007) 385-390.

[9] G. Bahrami, B. Mohammdi, J. Chromatogr. B 850 (2006) 417-422.

[10] Y. Jiang, J. Wang, H. Li, Y. Wang, J. Gu, J. Pharm. Biomed. Anal. 43 (2007) 1460-1464.

[11] X. Peng, Z. Wang, J. Li, G. Le, Y. Shi, Anal. Lett. 41 (2008) 1184-1199.

[12] J. Shin, D. F. Pauly, J. A. Johnson, R. F. Frye, J. Chromatogr. B 871 (2008) 130-134.

[13] X. Lu, L. Chen, D. Wang, J. Liu, Y. Wang, F Li, Chromatographia 68 (2008) 617-622.

[14] J. S. Torano, H. J. Guchelaar, J. Chromatogr. B 720 (1998) 89-97.

[15] A. Pappa-Louisi, A. Papageorgiou, A. Zitrou, S. Sotiropoulos, E. Georgarakis, F. Zougrou, J. Chromatogr. B 755 (2001) 57-64.

[16] J. I. D. Wibawa, P. N. Shaw, D. A. Barrett, J. Chromatogr. B 783 (2003) 359-366.

[17] S. Gurule, P. R. P. Verma, T. Monif, A. Khuroo, P. Partani, J Liq. Chromatogr. & Relat. Technol. 31 (2008) 2955-2973.

[18] F. D. Velde, J. W. C. Alffenaar, A, M. A. Wessels, Ben Greijdanus, D. R. A. Uges, J. Chromatogr. B 877 (2009) 1771-1777.

[19] S. Ostwald, J. Peters, M. Venner, W. Siegmund, J. Pharm. Biomed. Anal. 55 (2011) 194-201.

[20] D.H. Vu, R.A. Koster, M.S. Bolhuis, B. Greijdanus, R.V. Altena, D.H. Nguyen, J.R.B.J. Brouwers, D.R.A. Uges, J.W.C. Alffenaar, Talanta 121 (2014) 9-17.

[21] Guidance for Industry, Bioanalytical Method Validation, US Department of Health and Human Services, Food and Drug Administration Centre for Drug Evaluation and Research (CDER). Centre for Veterinary Medicine (CVM), May 2001.

[22] Guidance for Industry, ICH E6 Good Clinical Practice, U.S. Department of Health and Human Services, Food and Drug

Administration, Centre for Drug Evaluation and Research (CDER), Centre for Biologics Evaluation and Research (CBER), 1996.

[23] M. Yadav, P. S. Shrivastav, Bioanalysis 9 (2011) 1007-1024.

CHAPTER – 5

Sensitive and rapid determination of gliclazide in human plasma by UPLC-MS/MS and its application to a bioequivalence study

☞ *CONTENTS*

5.1 PREAMBLE

5.2 INTRODUCTION

5.3 EXPERIMENTAL AND METHOD OPTIMIZATION

5.3.1 Liquid chromatographic and mass spectrometric conditions

5.3.2 Preparation of calibration standards and quality control samples

5.3.3 Extraction procedure

5.3.4 Validation methodology

5.4 RESULTS AND DISCUSSION

5.4.1 Bioanalytical method development

5.4.2 Method validation results

5.5 BIOEQUIVALENCE STUDY DESIGN AND INCURRED SAMPLE REANALYSIS

5.6 CONCLUSION

5.7 REFERENCES

5.1 Preamble

An UPLC-MS/MS method has been developed for the determination of gliclazide in human plasma using gliclazide-d4 as the internal standard. The plasma samples were prepared by protein precipitation with acetonitrile employing 50 µL human plasma. Chromatography was performed on Acquity UPLC BEH C18 (50 mm × 2.1 mm, 1.7 µm) analytical column under isocratic conditions using a mobile

105

phase which consisted of 0.1% formic acid in water-acetonitrile (10:90, *v/v*). The MRM transitions for gliclazide (*m/z* 324.2 → 127.3) and gliclazide-d4 (*m/z* 328.2 → 127.4) were monitored on a triple quadrupole mass spectrometer, operating in the positive ionization mode. The method was validated over a dynamic concentration range of 1.0-2000 ng/mL for gliclazide. Matrix effect was assessed by post-column analyte infusion and the mean extraction recovery was 95.7 % across six quality control levels. Stability of gliclazide in plasma was evaluated under different conditions like, bench top, auto sampler, dry and wet extract, freeze-thaw and long term stability. The method was applied to a bioequivalence study with 30 mg gliclazide tablet formulation in 28 healthy subjects under fasting. Further, the assay reproducibility was confirmed by reanalysis of 129 incurred samples.

Keywords: Gliclazide; UPLC-MS/MS; protein precipitation; sensitive; bioequivalence study

5.2 Introduction

Diabetes mellitus is a metabolic disorder characterized by disrupted insulin production, leading to increase in blood glucose level and other complications such as neuropathy, cardiopathy and renal dysfunction [1-3]. It is classified as type 1 or type 2 diabetes based on different stages of the disease. Type 1 diabetes is ascribed to an early-onset autoimmune disease marked by the destruction of β-cells of the pancreas, which results in a partial or complete lack of insulin production and the inability of the body to control glucose homeostasis [3]. Type 2, a non-insulin dependent diabetes is a complex metabolic disorder which occurs at a later stage and is most common in the overweight population [4]. It is caused by genetic and environmental factors, as evident from a study correlating the loss of function gene variants in *GPR120* with increased risk of type 2 complications [5, 6]. Gliclazide is a potent type 2 second generation antidiabetic drug used to enhance insulin secretion, and possesses beneficial extra pancreatic effects that makes it potentially useful in type 1 as well [7]. Gliclazide is completely absorbed from the gastro-intestinal tract with mean absolute bioavailability of 97 % and is highly protein bound. After oral administration the peak plasma concentration is achieved in about 6 h and is more than 90 % recovered unchanged in plasma. Although six main metabolites are identified in urine, no active metabolites are found in plasma [8].

Either monotherapy or combination therapy is required for maintaining long term glycemic control with oral antidiabetic agents. Further, it is essential to monitor their plasma concentration for effective control of blood glucose levels and for therapeutic drug monitoring to optimize dose strength and dosing regimen. Thus, development of adequately sensitive, selective and rapid methods is required to determine these agents in plasma. Several methods are reported to determine gliclazide, either as a single analyte [9-14] or in combination with other antidiabetic agents [15-21] in human serum or plasma. These methods have employed high performance liquid chromatography with UV [9, 11, 14, 17, 19], electrochemical [12] or mass detection [10, 13, 15, 16, 18, 20, 21] for estimation of gliclazide. However, there are no reports on the use of UPLC-MS/MS for the quantification of gliclazide in human plasma. Thus, in the present study a selective, sensitive and rapid UPLC-MS/MS method has been developed and validated for accurate determination of gliclazide in human plasma for routine therapeutic drug monitoring. The method employs 50 μL plasma sample and a chromatographic analysis time of 1.2 min. The effect of endogenous matrix components on the quantification of gliclazide and its stability in plasma is extensively studied. Further, the utility of the method is demonstrated by a bioequivalence study in healthy subjects.

5.3   Experimental and method optimization

5.3.1 Liquid chromatographic and mass spectrometric conditions

An Acquity UPLC system from Waters Corporation (Milford, MA, USA) consisting of binary solvent manager, sample manager and column manager and linked to a triple quadrupole Quattro Premier XE mass spectrometer with an electrospray ionization source (Milford, MA, USA) was used in the study. The analysis of gliclazide and IS was performed on a Waters Acquity UPLC BEH C18 (50 mm × 2.1 mm, 1.7 μm) analytical column, maintained at 25 °C in a column oven. The mobile phase consisted of 0.1% formic acid in water - acetonitrile (10:90, *v/v*). The flow rate of the mobile phase was kept at 0.300 mL/min. The sample manager temperature was maintained at 5°C and the pressure of the system was 6200 psi.

Ionization and detection of gliclazide and IS was carried out on a triple quadrupole mass spectrometer in the positive ionization mode. The source dependent parameters for the analyte and IS were, cone gas flow: 120 L/h; desolvation gas flow: 640 L/h; capillary voltage: 1.4 kV, source temperature: 120°C; desolvation temperature: 400°C; extractor voltage: 4.0V. The pressure of argon used as collision activation dissociation gas was 0.120 Pa. The optimum value for cone voltage and collision energy was kept at 25 V and 31 eV for gliclazide and 24 V and 32 eV for IS respectively. Quadrupole 1 and 3 were maintained at unit mass resolution and the dwell time was set at 100 ms. Mass Lynx software version 4.1 was used to control all parameters of UPLC and MS.

5.3.2 *Preparation of calibration standards and quality control samples*

The standard stock solution of gliclazide (500 μg/mL) was prepared by dissolving requisite amount in methanol. Further, an intermediate solution (100 μg/mL and 50.0 μg/mL) for spiking was prepared in methanol:water (50:50, *v/v*). Calibration standards and quality control (QC) samples were prepared by spiking blank plasma with the intermediate solutions. Calibration curve standards for gliclazide were prepared at 1.00, 2.00, 5.00, 15.0, 45.0, 100, 250, 500, 1000, 2000 ng/mL concentrations, while quality control samples were prepared at six levels, 1600 ng/mL (HQC, high quality control), 1200/800 ng/mL (MQC-1/2, medium quality control), 400/3.00 ng/mL (LQC-1/2, low quality control) and 1.00 ng/mL (LLOQ QC, lower limit of quantification quality control). Stock solution (100 µg/mL) of the internal standard was prepared by dissolving 1.0 mg of gliclazide-d4 in 10.0 mL of methanol. Its working solution (2.50 µg/mL) was prepared by appropriate dilution of the stock solution in methanol:water (50:50, *v/v*). All standard stock and working solutions used for spiking were stored at 5 °C, while CSs and QC samples in plasma were kept at -70 °C until use.

5.3.3 Extraction procedure

Prior to analysis, all frozen subject samples, calibration standards and quality control samples were thawed and allowed to equilibrate at room temperature. To an aliquot of 50 µL of spiked plasma sample/subject sample, 25 µL of internal standard was added and vortexed for 10s. Further, 600 µL of acetonitrile was added and vortexed for another 60 s. The samples were then centrifuged at 13148 × g for 10 min at 10°C. The supernatant layer was separated and evaporated to dryness in a thermostatically controlled water-bath maintained at 40 °C under a gentle stream of nitrogen for 10 min. After drying, the residue was reconstituted with 250 µL of mobile phase solution. The solution was briefly vortexed for 15s and 10 µL was used for injection in the chromatographic system.

5.3.4 Validation methodology

The method validation was performed as per the USFDA guidelines [22]. Details of validation procedure and acceptance criteria are given in *Chapter-2*.

5.4 Results and discussion

5.4.1 Bioanalytical method development

In the present work electrospray ionization source was used to maximize sensitivity and obtain good linearity in the regression curves. During ionization in the positive ionization mode, gliclazide and IS formed predominantly protonated precursor ions in the full scan Q1 mass spectra at *m/z* 324.2 and 328.2 respectively. The most abundant product ions in Q3 mass spectra were found at *m/z* 127.3 and 127.4 for gliclazide and its deuterated analog as IS but applying optimum collision energy of 31 and 32 eV respectively *(Figure 1)*. This product ion can be attributed to the substructure containing hexahydro-cyclopenta[c]pyrrole moiety. In addition to the quantification transition, a qualifying transition was also monitored for the identification of gliclazide (*m/z* 324.2 → 110.2) and IS (*m/z* 328.2 → 110.4). A dwell time of 100 ms was adequate to have sufficient no. of data points for quantification.

Reported methods have used several reversed-phase columns like Techsphere C8 [9], Hypersil BDS C18 [10, 15], Cepcell Pak C18 [11], Apollo C18 [12], Diamonsil C18 [13] with different dimensions for chromatographic separation of gliclazide. Foroutan *et al*. [14] have used a monolithic column for the analysis of gliclazide from human plasma. In the present work, the chromatographic analysis of gliclazide was suitably optimized on UPLC BEH C18 (50 mm × 2.1 mm, 1.7 μm) column using various combinations of acetonitrile/methanol with acidic buffers (formic acid-ammonium formate and acetic acid-ammonium acetate) and acidic modifiers like formic acid and acetic acid in different concentrations. The aim was to have adequate retention, response and peak shape with a short run time.

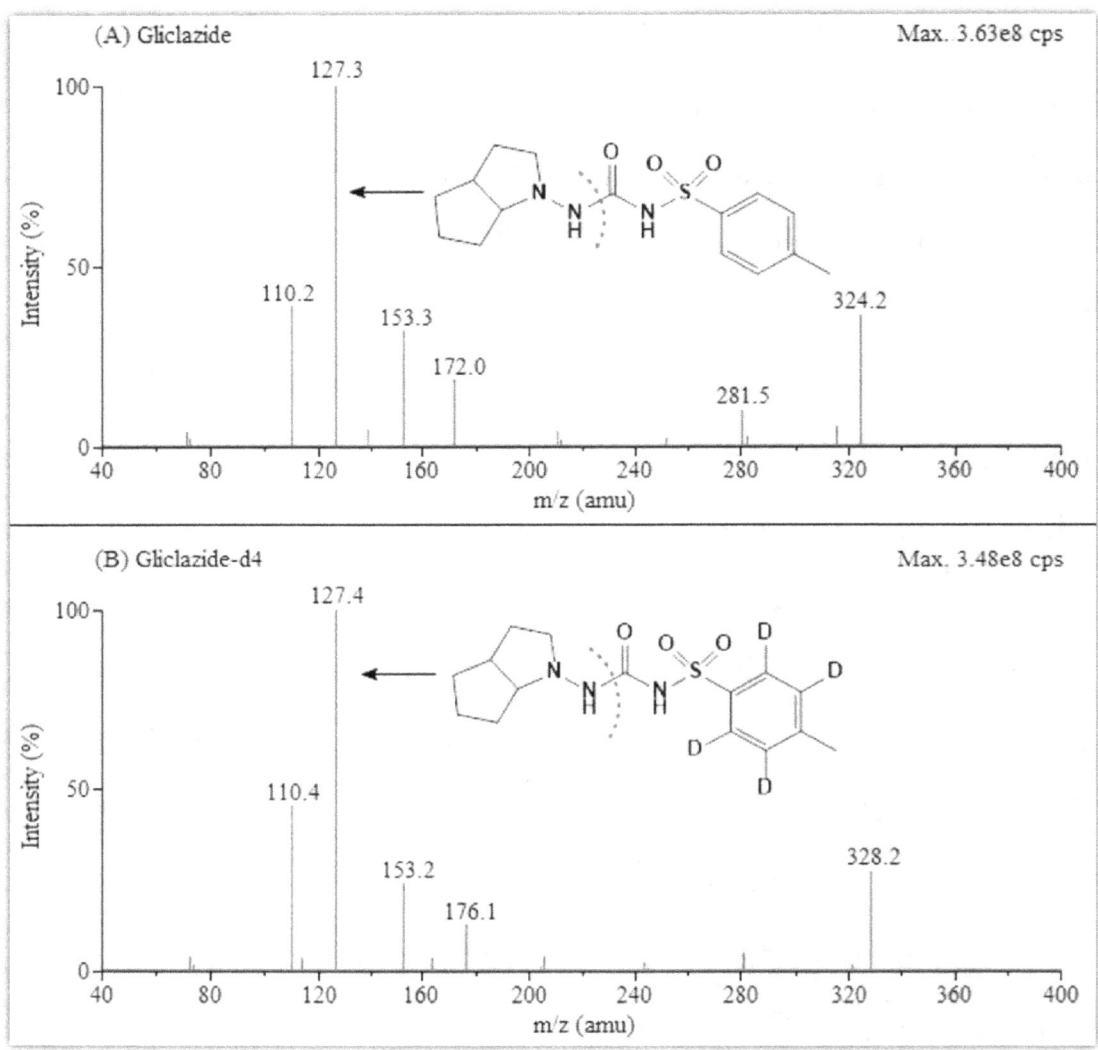

Figure 1.Product ion mass spectra of (A) gliclazide (m/z 324.2 → 127.3, scan range 40-400 amu) and (B) internal standard, gliclazide-d4 (m/z 328.2 → 127.4, scan range 40-400 amu) in the positive ionization mode.

Although adequate retention was obtained under all mobile phase conditions, the peak shape was not acceptable when the aqueous part was greater than 25 % with longer retention time. Further, the response was much higher in formic acid compared to both the acidic buffers. Thus, a mobile phase consisting of 0.1% formic acid in water - acetonitrile (10:90, *v/v*) was found best in terms of sensitivity, peak shape and short analysis time. The entire chromatographic run was completed within 1.2 min. All previous methods [9-16] have used general internal standards (ISs) for the analysis of gliclazide. However, according to the FDA guidelines [22] a stable and labeled compound should be preferred which has more structural similarity compared to general standards. Gliclazide-d4, used in this work gave excellent results with acceptable accuracy and precision at each QC level. The MRM chromatograms illustrated in *Figure 2* of extracted blank plasma (double blank), blank plasma fortified with IS, gliclazide (at 1.0 ng/mL) and IS and a subject sample at C_{max} demonstrates the overall performance and selectivity of the method. Results of post-column infusion experiment in *Figure 3* indicate no ion suppression or enhancement at the retention time of gliclazide and IS.

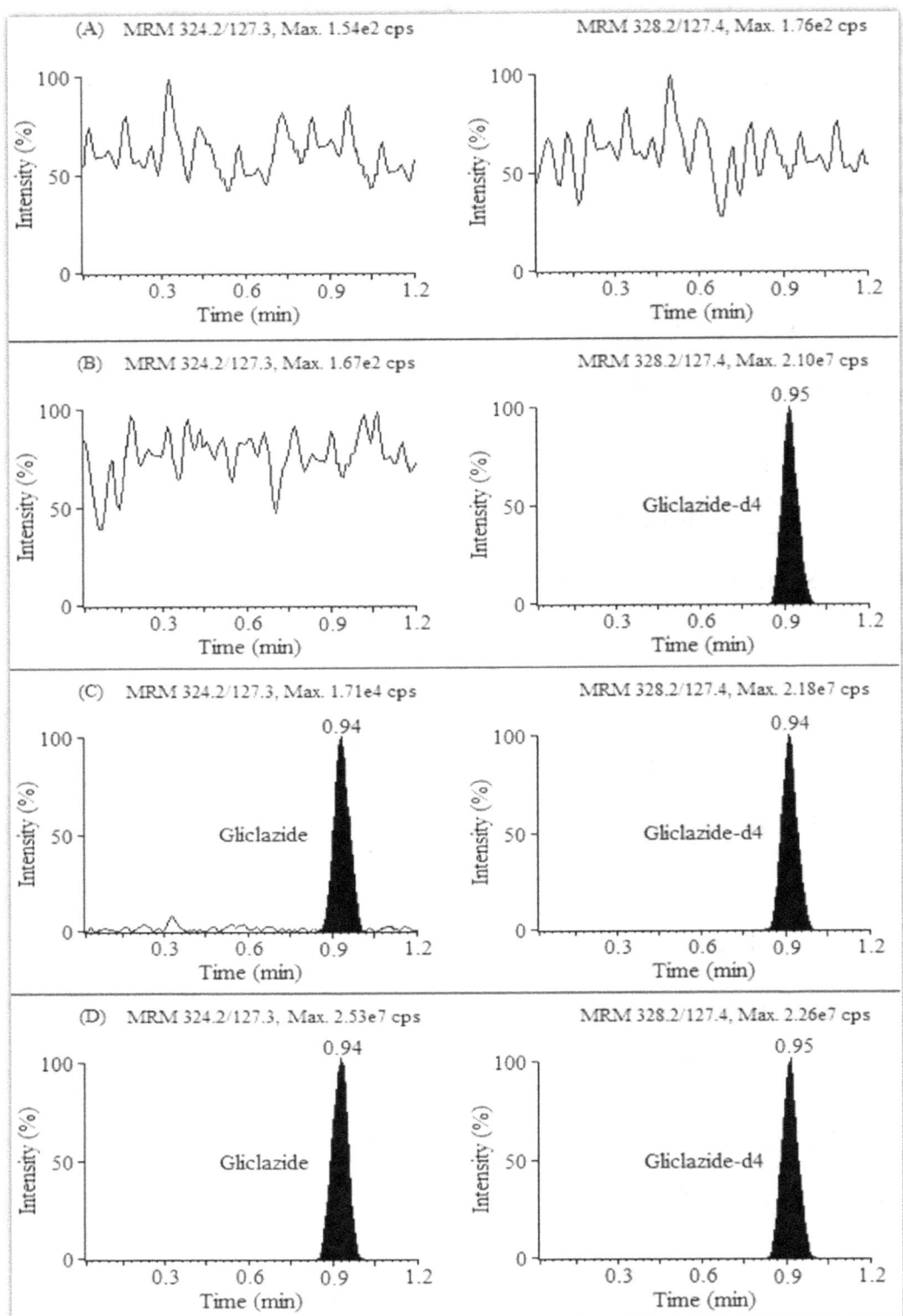

*Figure 2.***Representative MRM ion-chromatograms of (A) double blank plasma without gliclazide and gliclazide-d4, (B) blank plasma with working solution of gliclazide-d4 (m/z 328.2 → 127.4), (C) gliclazide (m/z 324.2 → 127.3) at 1.0 ng/mL concentration of gliclazide and gliclazide-d4 (D) real subject sample at Cmax and gliclazide-d4 after administration of 30 mg gliclazide.**

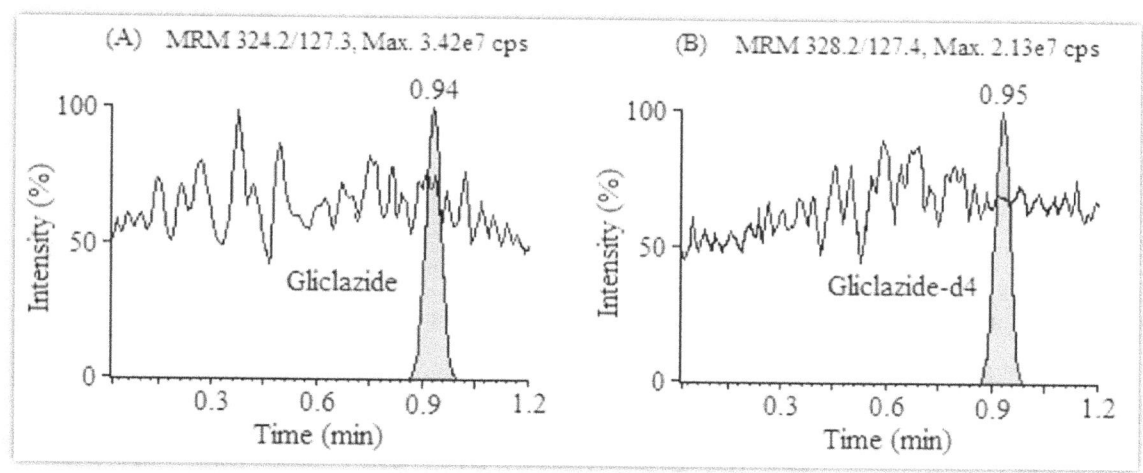

Figure 3.Injection of extracted blank human plasma during post column infusion of (A) gliclazide at 2000 ng/mL and (B) gliclazide-d4.

Reported methods have used either protein precipitation (PP) [12, 14-16] or liquid-liquid extraction (LLE) [9, 11, 13] for sample preparation of gliclazide from plasma or serum samples. However, to keep the extraction protocol simple and quick with minimum steps, PP was tried with different protein precipitants like acetonitrile, methanol, ethanol and acetone. The extraction efficiency obtained in these solvents was quantitative ($\geq$ 81 %), but minimum interference of endogenous components was found in presence of acetonitrile. At the same time the peak shape and analyte response was much higher and consistent at all QC levels with acetonitrile and hence was selected in the present work.The salient features of the present work in comparison with reported methods are presented in ***Table 1***.

5.4.2 Method validation results

The calibration curves were linear over the concentration range of 1.0-2000 ng/mLwith correlation coefficient (r^2) $\geq$ 0.9995 for gliclazide. The mean equation for five calibration curves was $y = (0.00081 \pm 0.00001) \, x - (0.000041 \pm 0.000034)$. The standard deviation value for slope, intercept and correlation coefficient observed were 0.00001, 0.000034 and 0.0001 respectively. The accuracy and precision (%CV) observed for the calibration curve standards ranged from 97.6 to 101.6 % and 0.88 to 2.92 % respectively. The lowest concentration (1.00 ng/mL) in the standard curve was measured at a signal-to-noise ratio (S/N) of $\geq$ 10.0.

Ref.	Technique	Sample volume (μL); extraction procedure; internal standard	Linear range (ng/mL); retention time /run time (min)	Application
		Table 1. Comparison of salient features of chromatographic methods developed for gliclazide in biological matrices		
9	HPLC-UV (230 nm)	100 (serum); LLE with toluene; phenytoin	75-10,000; 6.8/8.0	Pharmacokinetic study with 80 mg gliclazide in 12 healthy volunteers
10	LC-MS/MS	1000 (plasma); Non-porous membrane probe; tolbutamide	100-9850; 2.71/3.00	Pharmacokinetic study with 80 mg gliclazide in 1 healthy volunteer
11	Semi-micro HPLC-UV (229 nm)	100 (plasma); LLE with chloroform; glyburide	100-10,000; 4.1/8.0	Bioequivalence study with two formulations in 20 healthy subjects
12	HPLC-electrochemical detection	100 (plasma); PP with ACN; methyl-4-hydroxy benzoate	16-1300; 5.0/20.0	Pharmacokinetic study with 80 mg gliclazide in 1 healthy volunteer
13	LC-MS/MS	200 (plasma); LLE with *n*-hexane and DCM; tolbutamide	2.5-2000; 3.55/5.0	Pharmacokinetic study with 30 mg gliclazide in 20 patients
14	HPLC-UV (230 nm)	450 (plasma); PP with ACN; glibenclamide	10-5000; 3.2/6.0	Pharmacokinetic study with 80 mg gliclazide in 12 healthy volunteer
15[a]	LC-MS/MS	200 (plasma); PP with ACN; huperzine	10-10000; 1.47/2.0	Bioequivalence study with 40 mg gliclazide in 20 healthy subjects
16[b]	LC-MS/MS	100 (plasma); PP with methanol; ketoconazole	20-2000; 6.5/8.0	Protein binding interaction with Rhein
PW	UPLC-MS/MS	50 (plasma); PP with ACN; gliclazide-d4	1.0-2000; 0.94/1.20	Bioequivalence study with 30 mg gliclazide in 28 healthy subjects

[a]Alongwith metformin; [b]Along with benazepril and valsartan; LLE: liquid-liquid extraction; PP: protein precipitation; ACN: acetonitrile; DCM: dichloromethane

The intra-batch and inter-batch precision and accuracy results obtained at each QC level are shown in *Table 2*. The intra-batch precision (%CV) ranged from 1.04 to 2.89 % and the accuracy was within 99.6 to 100.7 %. For the inter-batch experiments, the precision varied from 1.09 to 2.95 % and the accuracy ranged from 98.6 to 100.4 %.

The extraction recovery and matrix effect results for gliclazide and IS at different QC levels are presented in *Table 3*. The mean recovery across QC levels for gliclazide and IS ranged from 94.6 to 96.7 %, while the matrix effect expressed matrix factors varied from 0.983 to 1.014. Further, another essential criteria to evaluate matrix effect, expressed as relative matrix in different plasma lots/batches showed % CV values in the measurement of slope of standard curves was 2.86 *(Table 4)*.

Table 2. Intra-batch and inter-batch precision and accuracy for Gliclazide						
Nominal concentration (ng/mL)	Intra-batch (n = 6; single batch)			Inter-batch (n = 30; 6 from each batch)		
	Mean conc. found (ng/mL)	% CV	% Accuracy	Mean conc. found (ng/mL)	% CV	% Accuracy
HQC (1600)	1604.2	2.07	100.2	1591.0	2.23	99.4
MQC-1 (1200)	1195.0	1.04	100.3	1204.6	1.09	99.6
MQC-2 (800.0)	802.44	2.41	99.6	797.24	2.95	100.4
LQC-1 (400.0)	402.48	2.37	100.6	398.14	1.92	99.5
LQC-2 (3.00)	3.018	2.89	100.6	2.987	2.62	99.5

LLOQ QC (1.000)	1.007		2.81	100.7	0.986	2.91 98.6

Table 3. Extraction recovery and matrix factor for gliclazide at different QC levels

QC conc. (ng/mL)	Mean area response (n = 6)			Extraction recovery, % (B/A)	Matrix factor		
	A	B	C		Analyte (A/C)	IS	IS-normalized
3.00	7292	6935	6718	95.1 (96.3)[a]	1.085	1.070	1.014
400.0	974865	935241	904587	95.9 (95.9)[a]	1.077	1.083	0.994
800.0	1985678	1878746	1848514	94.6 (96.7)[a]	1.074	1.062	1.011
1200	2928627	2816242	2737784	96.2 (95.1)[a]	1.069	1.079	0.990
1600	3895563	3762278	3688412	96.6 (94.7)[a]	1.056	1.074	0.983

[a]values for gliclazide-d4;

A: mean area response of six replicates prepared by spiking in extracted blank plasma;

B: mean area response of six replicates prepared by spiking before extraction;

C: mean area response of six replicates prepared by spiking in mobile phase (neat samples)

Table 4. Relative matrix effect in eight different lots of human

plasma for Gliclazide	
Plasma lot	**Slope**
Lot-1	0.000803
Lot-2	0.000792
Lot-3	0.000821
Lot-4	0.000845
Lot-5	0.000788
Lot-6 (heparinized)	0.000817
Lot-7 (haemolysed)	0.000832
Lot-8 (lipemic)	0.000778
Mean	0.000810
±SD	0.000023
%CV	2.84

The stock solutions for short term stability of gliclazide and IS were stable at room temperature up to 27 h and between 2-8 °C for a minimum period of 24 days. Gliclazide in control human plasma (bench top) at room temperature was stable for at least 20 h at 25°C and for minimum of five freeze and thaw cycles. Dry and wet extract stability of the spiked quality control samples was determined up to 24 h and 48 h respectively. Long term stability of the spiked quality control samples was remained unaffected up to 186 days. The detailed stability results at LQC and HQC levels in plasma are presented in *Table 5*.

Table 5. Stability results of gliclazide in plasma under various conditions (n = 6)			
Storage conditions	**Nominal concentrat**	**Mean stability**	**Change (%)**

	ion (ng/mL)	sample (ng/mL) ± SD	
Bench top stability at 25 °C, 20 h	1600	1604.4 ± 39.8	0.28
	3.00	2.99 ± 0.07	-0.19
Freeze & thaw stability at -20 °C	1600	1606.0 ± 35.8	0.38
	3.00	3.019 ± 0.06	0.65
Freeze & thaw stability at -70 °C	1600	1594.0 ± 36.0	-0.38
	3.00	2.99 ± 0.07	-0.22
Dry extract stability at 2-8°C, 24 h	1600	1598.6 ± 35.7	-0.09
	3.00	3.02 ± 0.07	0.51
Wet extract stability at 2-8°C, 48 h	1600	1608.2 ± 31.5	0.51
	3.00	2.98 ± 0.06	-0.63
Long term stability at -20 °C, 186 days	1600	1609.4 ± 30.5	0.59
	3.00	3.01 ± 0.08	0.33
Long term stability at -70 °C, 186 days	1600	1607.4 ± 19.8	0.46
	3.00	2.98 ± 0.06	-0.34

For method ruggedness, the precision (%CV) and accuracy values for two different columns ranged from 2.13 to 4.45 % and 97.6 to 104.6 % respectively across five quality control levels. For the experiment with different analysts, the results for precision and accuracy were within 1.29 to 2.05 % and 95.1 to 102.4 % respectively at these levels. The dilution integrity experiment was performed with an aim to validate the dilution test to be carried out on higher analyte concentration above the upper limit of quantification (ULOQ), which may be encountered during real subject sample analysis.The precision and accuracy values for 1/2th and 1/10th dilution ranged from 2.95-3.46 % and 97.8-105.4 % for gliclazide.

5.5　*Bioequivalence study design and incurred sample reanalysis*

The bioequivalence study was conducted with a single dose of a modified release tablets (30 mg gliclazide, Valpharma International S.p.A., Italy) and DIAMICRON® modified release tablet (30 mg of gliclazide, Les Laboratories Servier, France)formulations to 28 normal, healthy, adult subjects under fasting. The study was conducted as per the International Conference on Harmonization, E6 Good Clinical Practice guidelines [23]. Blood samples were collected at 0.00 (pre-dose), 0.50, 1.00, 1.50, 2.00, 2.50, 3.00, 3.50, 4.00, 4.50, 5.00, 5.50, 6.00, 6.50, 7.00, 7.50, 8.00, 9.00, 10.0, 11.0, 12.0, 14.0, 16.0, 24.0, 48.0, 72.0, 96.0 and 120 h after oral administration of the dose for test and reference formulation. *Figure 4* shows the plasma concentration vs. time profile of gliclazide in healthy subjects under fasting.

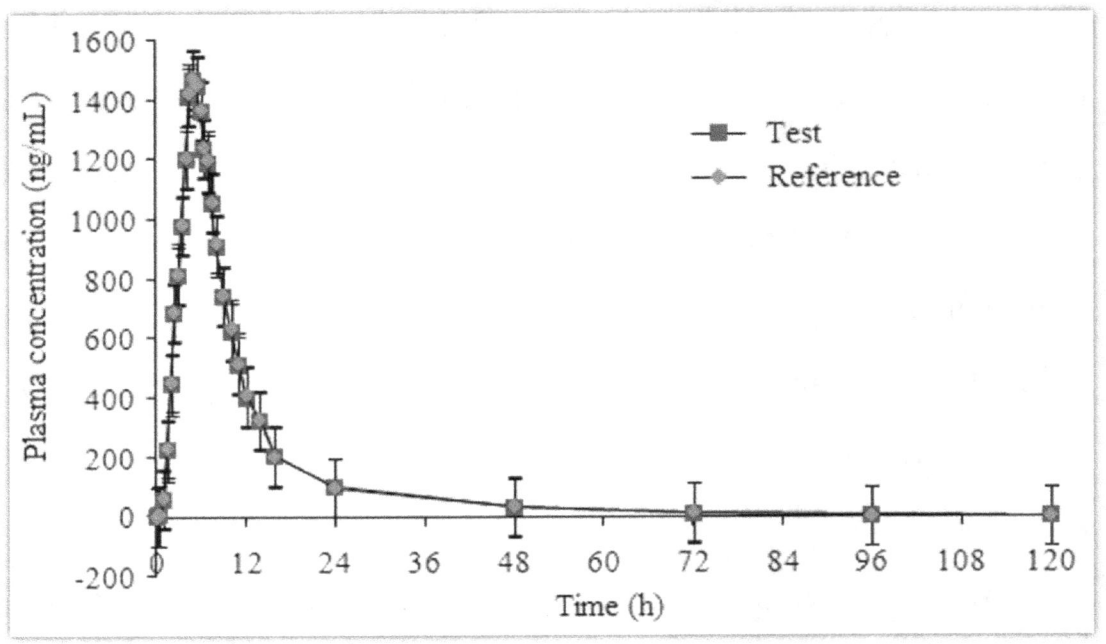

Figure 4. **Mean plasma concentration-time profile of gliclazide after oral administration of 30 mg of test (modified release tablet, Valpharma International, Italy) and reference (DIAMICRON® tablet, Les Laboratories Servier, France) formulation to 28 healthy subjects.**

It was possible to analyze about 2500 samples (subject samples, calibration and QC samples) during a period of 8 days and the results obtained were well within the acceptable limits. The main pharmacokinetic parameters of gliclazide, C_{max}, AUC_{0-t}, AUC_{0-inf}, T_{max}, K_{el} and $t_{1/2}$ for both the formulations are shown in **Table 6**. The C_{max}, T_{max}, K_{el} and $t_{1/2}$ values obtained were in good agreement with a previous report with 30 mg gliclazide modified release tablets in Chinese subjects [13]. Further, there was no apparent statistical differencebetween the two formulations in any parameter. The ratios of mean log-transformed parameters (C_{max}, AUC_{0-120}, and AUC_{0-inf}) and their 90% CIs were all within the defined bioequivalence range of 80-125 %. These observations confirm the bioequivalence of the test formulation with the reference product in terms of rate and extent of absorption.

Table 6. Mean pharmacokinetic parameters, comparison of treatment ratios of 90% CIs of natural log (Ln)-transformed parameters following oral administration of 30 mg gliclazide tablet formulation in 28 healthy Indian subjects under fasting						
Parameter	*Test (Mean ±SD)*	*Reference (Mean ±SD)*	*Ratio (test/ reference), %*	*90% CI (Lower – Upper)*	*Power*	*Intra subject variation, % CV*
C_{max} (ng/mL)	1465.52 ± 190.25	1472.31 ± 205.36	99.5	95.3-104.1	0.9995	4.65
$AUC_{0-120 h}$ (h. ng/mL)	11245.2 ± 2045.5	12045.6 ± 2060.9	99.3	93.1-7	0.9997	5.25
AUC_{0-inf} (h. ng/mL)	11965.4 ± 2102.6	12445.0 ± 2132.6	96.1	92.7-99.5	0.9998	4.92
T_{max} (h)	5.05 ± 1.25	5.26 ± 1.35	--	--	--	--
$t_{1/2}$ (h)	9.19 ± 1.24	9.46 ± 1.02	--	--	--	--

Kel	0.075 ±	0.073 ±				
(1/h)	0.006	0.005	--	--	--	--

The incurred sample reanalysis (ISR) results are represented in *Figure 5*. The % change for assay reproducibility in 129 incurred samples was within ±16 % for gliclazide, which is within the acceptance criterion of ± 20 % [24]. This authenticates the reproducibility of the proposed method.

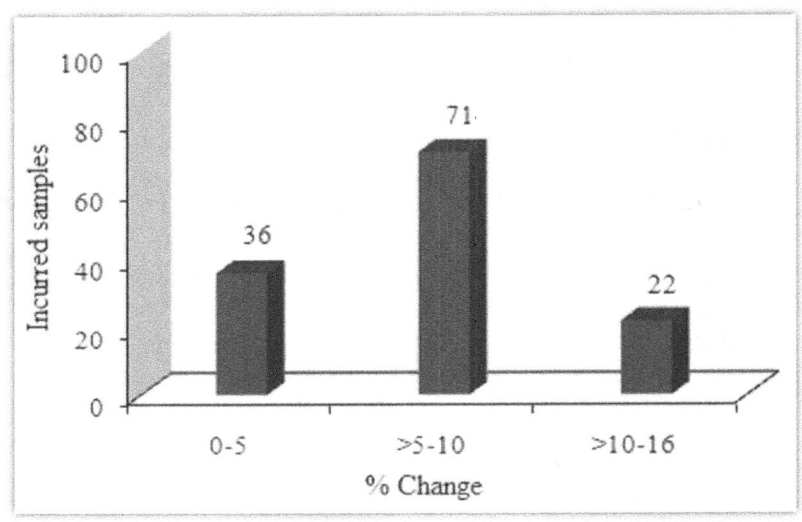

Figure 5.Graphical representation of % change in concentration during reanalysis of 129 incurred samples for gliclazide.

5.6 Conclusion

The validated UPLC-MS/MS method offers several advantages over reported procedures, in terms of lower sample requirements, simplicity of extraction procedure and overall analysis time. The proposed method is more sensitive and rapid (chromatographically) compared to all other procedures for determination of gliclazide in biological fluids. The present method employs small plasma volume (50 µL) for processing, which is lower compared to all existing methods for gliclazide. With dilution integrity up to two folds, it is possible to extend the upper limit of quantification to 4000 ng/mL. Matrix effect is assessed through post-column analyte infusion and precision values for calculated slopes of calibration curves. The validated method showed acceptable study data for the quantification of gliclazide in a clinical setting. Further, incurred sample reanalysis of 129 samples authenticates the reproducibility of the proposed method and is reported in any previous method.

5.7 References

[1] J.F. Yale, J. Am. Soc. Nephrol. 16 (2005) S7-S10.

[2] H.E. Lebovitz, Med. Clin. North Am. 88 (2004) 847-863.

[3] A. Mooranian, R. Negrulj, N. Chen-Tan, H. S. Al-Sallami, Z. Fang, T. Mukkur, M. Mikov, S. Golocorbin-Kon, M. Fakhoury, F. Arfuso, H. Al-Salami, Drug Design Dev. Ther. 8 (2014) 1003-1012.

[4] E. Ferrannini, Endocr. Rev. 19 (1998) 477-490.

[5] D.Y. Oh, J.M. Olefsk, Cell Metab. 15 (2012) 564-565.

[6] R.N. Burns, N.H. Moniri, Biochem. Biophys. Res. Commun. 396 (2010) 1030-1035.

[7] H. Al-Salami, G. Butt, I. Tucker, R. Skrbic, S. Golocorbin-Kon, M. Mikov, Med. Hypothesis Res. 4 (2008) 93-101.

[8] DIAMICRON® MR (gliclazide) Product Monograph, June 2012. Servier Canada Inc.235 Boulevard Armand Frappier, Laval, Quebec H7V 4A7, Canada. http://www.servier.ca/sites/default/files/webform/Products/EN-DIAMICRON-MR-PI.pdf

[9] M.R. Rouini, A. Mohajer, M.H. Tahami, J. Chromatogr. B 785 (2003) 383-386.

[10] F.G.P. Mullins, J. Sep. Sci. 24 (2001) 593-598.

[11] J.Y. Park, K.A. Kim, S.L. Kim, P.W. Park, J. Pharm. Biomed. Anal. 35 (2004) 943-949.

[12] C.Y. Kuo, S.M. Wu, J. Chromatogr. A 1088 (2005) 131-135.

[13] G. Ling, J. Sun, J. Tang, X. Xu, Y. Sun, Z. He, Anal. Lett. 39 (2006) 1381-1391.

[14] S.M. Foroutan, A. Zarghi, A. Shafaati, A. Khoddam, J. Pharm. Biomed. Anal. 42 (2006) 513-516.

[15] G.P. Zhong, H.C. Bi, S. Zhou, X. Chen, M. Huang, J. Mass Spectrom. 40 (2005) 1462-1471.

[16] X. Hu, Y. Zheng, J. Sun, L. Shang, G. Wang, H. Zhang, Chromatographia 69 (2009) 843-852.

[17] S. AbuRuz, J. Millership, J. McElnay, J. Chromatogr. B 817 (2007) 277-286.

[18] K. Abro, N. Memon, M.I. Bhanger, S. Perveen, A. Panhwar, Anal. Lett. 45 (2012) 1947-1959.

[19] P. Venkatesh, T. Harisudhan, H. Choudhary, R. Mullangi, N.R. Srinivas, Biomed. Chromatogr. 20 (2006) 1043-1048.

[20] H.H. Maurer, C. Kratzsch, T. Kraemer, F.T. Peters, A.A. Weber, J. Chromatogr. B 773 (2002) 63-73.

[21] C. Hess, F. Musshoff, B. Madea, Anal. Bioanal. Chem. 400 (2011) 33-41.

[22] Guidance for Industry, Bionanlytical Method Validation, US Department of Health and Human Services, Food and Drug

Administration Centre for Drug Evaluation and Research (CDER), Centre for Veterinary Medicine (CVM), May 2001.

[23] Guidance for Industry: ICH E6 Good Clinical Practice, U.S. Department of Health and Human Services, Food and Drug Administration, Centre for Drug Evaluation and Research (CDER), Centre for Biologics Evaluation and Research (CBER), April 1996.

[24] M. Yadav, P.S Shrivastav, Bioanalysis 3 (2011) 1007-24